1993

University of St. Francis
GEN 510.1 R692
Rodriguez-Consuegra, Fra
The mathematical philosophy of

3 0301 00074859 6

D1561030

Francisco A. Rodríguez-Consuegra

The Mathematical Philosophy of Bertrand Russell: Origins and Development

College of St. Francis Library
Joliet, Illinois

1991

Birkhäuser Verlag
Basel · Boston · Berlin

Author's address

Prof. Dr. Francisco A. Rodríguez-Consuegra
Pere Martell, 7, 2° D
E- 43001 Tarragona, Spain

Deutsche Bibliothek Cataloging-in-Publication Data

Rodríguez-Consuegra, Francisco A.: The mathematical philosophy of Bertrand Russell: origins and development / Francisco A. Rodríguez-Consuegra. – Basel; Boston; Berlin: Birkhäuser, 1991
ISBN 3-7643-2656-5 (Basel ...) Gb.
ISBN 0-8176-2656-5 (Boston) Gb.

This work is subject to copyright. All rights are reserved, whether the whole or part of the material is concerned, specifically those of translation, reprinting, re-use of illustrations, broadcasting, reproduction by photocopying machine or similar means, and storage in data banks. Under § 54 of the German Copyright Law, where copies are made for other than private use a fee is payable to «Verwertungsgesellschaft Wort», Munich.

© 1991 Birkhäuser Verlag Basel
Printed from the author's camera-ready manuscripts on acid-free paper in Germany
ISBN 3-7643-2656-5
ISBN 0-8176-2656-5

To Ana,
who made my return to work possible

Contents

Keys to Russell's works and Cross references	x
Acknowledgements	xi
Preface by Ivor Grattan-Guinness	xiii
Introduction	1

1. Methodological and logicist background — 5

1.1.	Boole and Peirce	5
1.2.	Dedekind and Cantor	13
1.3.	Couturat and Whitehead	19
1.4.	Bradley and Moore	27
1.5.	*Foundations of geometry*	36

2. The unpublished mathematical philosophy: 1898-1900 — 44

2.1.	The genesis of the 1898-1900 manuscripts	44
2.2.	Logic, mathematics and ontology	49
2.3.	The evolution of the main concepts	57
2.4.	Concepts, axioms, presupposition and implication	62
2.5.	The contradiction and the infinite	69
2.6.	Relations and the 'principle of abstraction'	72
2.7.	The method of definition	77
2.8.	The gradual approach to Cantor	81
	2.8.1. The first contacts and opinions 81	
	2.8.2. The reasons for the rejection 86	

3. The contribution of Peano and his school — 91

3.1. Logic — 92
3.1.1. Objective and stages 92
3.1.2. Primitives, logical order and interdefinability 93
3.1.3. Implication, inclusion and membership 97
3.1.4. Classes, propositions and individuals 99
3.1.5. Mathematical propositions and quantification 101
3.1.6. Relations, functions, classes, properties and propositions 103

3.2. Arithmetic — 105
3.2.1. The axioms and their interpretation: Dedekind 105
3.2.2. The definability of number 107
3.2.3. Real numbers: construction and definition 108
3.2.4. The 'logicist' arithmetic: Cantor 110

3.3. Geometry — 113
3.3.1. The geometric calculus and the principles of geometry 113
3.3.2. The 'logicist' geometry 117

3.4. The method — 119
3.4.1. Axiomatics 119
3.4.2. Definitions 121
3.4.3. The definition by abstraction 124
3.4.4. Simplicity, analysis and intuition 125

3.5. Peano's followers and their contributions — 127
3.5.1. The various improvements 127
3.5.2. The transformation of definitions by abstraction into nominal ones 131

4. The principles of mathematics — 135

4.1. The reaction to the Congress of 1900 — 135
4.1.1. The notes to the manuscripts: from Moore to Peano 135
4.1.2. The first writings 138
4.1.3. The acceptance of Cantor 141

4.2. Logic — 144
4.2.1. The indefinables and the propositional function 144
4.2.2. Relations 151

	4.3.	**Arithmetic**	155
		4.3.1. The definition of cardinal number 155	
		4.3.2. Finite and infinite 162	
		4.3.3. Quantity 164	
		4.3.4. Order 166	
		4.3.5. Ordinal numbers 169	
		4.3.6. Real numbers 173	
	4.4.	**Geometry**	175
	4.5.	**What Russell learned from Peano**	181

5. Philosophical and methodological problems 185

5.1. Origin and evolution of Russell's logicism 185
5.2. The principle of abstraction 189
5.2.1. Origin and evolution 189
5.2.2. Assumptions and implications 194
5.3. The constructive definition 205
5.3.1. Nominal definitions 205
5.3.2. Mathematical and philosophical definitions 208
5.3.3. Analysis and ordinary language 211
5.4. Relational logic and ontology 215

Bibliography 224

B.1. Works by Russell 224
B.1.1. Unpublished manuscripts 224
B.1.2. Published or unpublished correspondence 225
B.1.3. Published works 225
B.2. Works by other authors 227

Keys to Russell's works

I also include the bibliographical reference, the only key used for the rest of articles or books mentioned or quoted (see Bibliography).

AB	*Autobiography, 1967a*
AMR	*Analysis of mathematical reasoning, m1898*
CP	*The Collected Papers of Bertrand Russell, 1983a, 1984a...*
EA	*Essays in analysis, 1973a*
FG	*Foundations of geometry, 1897a*
FIAM	*Fundamental ideas and axioms of mathematics, m1899*
IMP	*Introduction to mathematical philosophy, 1919a*
LK	*Logic and knowledge, 1956a*
ML	*Mysticism and logic, 1918a*
MPD	*My philosophical development, 1959a*
OKEW	*Our knowledge of the external world, 1914a*
PE	*Philosophical essays, 1910b*
PL	*A critical exposition of the philosophy of Leibniz, 1900a*
PM	*Principia mathematica, 1910a, 1912a, 1913a*
POM1	*Principles of mathematics, m1900*
POM	*The principles of mathematics, 1903a*
PP	*The problems of philosophy, 1912b*
PRM	*Portraits from memory, 1956b*
TK	*Theory of knowledge* (1913), *1984a*

Cross references

All references, no matter whether denoting a chapter, a section or a sub-section, are always indicated through their numbers, i.e. 4, 4.5, 4.5.3.

Acknowledgements

I am very especially grateful to Ivor Grattan-Guinness, who encouraged me during the past years, helped me with several publications, read and improved the whole manuscript of this book through an impressive number of remarks, and wrote the Preface. I would like to express my thanks also to Jesús Mosterín for supervising the Ph. D. thesis which served as a partial ground for this book, and for supporting me during the composition. I am also grateful to Juan J. Acero, Alejandro Garciadiego, Mario Gómez, Javier de Lorenzo, Gregory Landini, Willard V. Quine, Josep Rifá and Roberto Torretti for reading some chapters (or related materials) and sending me valuable comments.

Concerning the unpublished materials which has been necessary to study, and all the related information during many years, the kind help of Kenneth Blackwell, the Russell Archivist, and the staff of *The Bertrand Russell Archives* (especially Carl Spadoni), McMaster University, Hamilton, Ontario, Canada, has been decisive, together with the permission of Christopher Farley (*The Bertrand Russell Estate*). The copyright of all the quoted unpublished material (manuscripts and correspondence) is held by McMaster University, so my thanks are also due to Louis Greenspan, Chairman of *The Bertrand Russell Archives Copyright Permissions Committee*, for his explicit permission.

Some parts of this book have been previously published. Thus, earlier Spanish versions of chapters 2, 3 and 4 have appeared as my *1988a, 1988b, 1988c* and *1988d*. Likewise, I have reproduced parts of my English paper *1987c*, mainly in 5.2. My thanks are due to both publishers (the Department of Mathematics of the National Autonomous University of Mexico — *Mathesis*— and Taylor & Francis, Ltd. —*History and Philosophy of Logic*) for their permission. However, the way in which these materials appear here is quite different, for I have taken advantage to make many corrections and improvements.

I am grateful to Benno Zimmermann, from Birkhäuser Verlag, for having accepted this book for publication, as well as to Doris Wörner for her efficient help with the details. Also, I would like to say that, although I myself typed the book camera-ready, the task was made much easier thanks to my Macintosh computer, as well as to the following software: Ms. Word, Expressionist and —in a few places— Hyper Card.

I cannot forget here the person of Bertrand Russell, whose works constituted the main reason I studied philosophy, after having finished another completely different degree. I still remember my frustration on reading, during a journey to Buenos Aires in 1970, the news of his death in *La Nación,* once I had decided to try to meet him some day. Michael Scott wrote (in his contribution to Schoenman *1967a*) that it is impossible to write about Russell without offending someone, perhaps including Russell himself. This idea, which I share, may be used to express my intention of regarding this book, no matter what its arguments or conclusions can state, as a little homage to him.

Cambrils, Tarragona, December 1990

After finishing this book, I received volume **8** of *Russell: the Journal of the Bertrand Russell Archives* (the proceedings of a conference on Russell's philosophy), and the books *Rereading Russell: Essays in Bertrand Russell's metaphysics and epistemology* (vol. XII of Minnesota Studies in the Philosophy of Science), and E.R. Eames, *Bertrand Russell's dialogue with his contemporaries*. I do not think that they contain ideas which can be regarded as relevant to the results of my investigations here; see my essays-review (*1990b*, *1990c* and *1991a* in the Bibliography.)

Preface

by Ivor Grattan-Guinness

Until twenty years ago the outline history of logicism was well known. Frege had had the important ideas, until he was eclipsed by Wittgenstein. Russell was important in publicising the former and tutoring the latter, and also for working with Moore in the conversion of British philosophy from neo-Hegelianism to the new analytic tradition in the 1900s, but his own work on logic and especially logicism was very muddled.

Around that time Russell, who was still alive, sold his manuscripts to McMaster University in Canada, and interest in his achievements in logic began to develop, especially after his death in 1970. Scholars found thousands of folios of unpublished holograph awaiting their attention, and also hundreds of pertinent letters (both in the Russell Archives and elsewhere in certain recipients' collections). Various facets of his work came to light for the first time, and others —which could have been gleaned from carefully reading of the published sources— gained new publicity from the evidence revealed in manuscripts. Even the technical passage work, which constitutes the unread majority of the *Principia mathematica* (1910-13) of Russell and Whitehead, began to receive a little respectful scrutiny.

It turned out that Russell had done several pioneering things. While indeed often incoherent in reference and content, they comprised major forays into the new mathematical logic, of which he turned out to be a major founder: some are even of interest to modern studies.

The historical 'record' also needed major rewriting. The rapid conversion of British philosophy from neo-Hegelianism never happened (instead, a long and slow change took place over some decades); even Russell himself retained elements of his old position, and to his benefit. For example, his conversion took place in several stages, for he tried several alternatives before the analytical paradise opened up for him in 1900. Again, his neo-Hegelian sensitivity to paradoxes in general must have helped to find his famous paradox of set theory early on his work (in contrast to the brainy Frege, who never spotted it, and was nearing the end of his toils when he learnt of the result from Russell).

Most of the new exegesis of Russell studies has been effected by English and North American scholars; but interest among other nations is becoming manifest —even among (a few) French, who normally have followed Russell's eminent contemporary Poincaré in ridiculing logic and its ism. Now a Spanish-native line is emerging. Already A. Garciadiego (Mexico) has given much attention to the history of all the paradoxes which Russell collected; and now we have the substantial fruits of the endeavours of F. A. Rodríguez-Consuegra (Tarragona).

In this book Rodríguez-Consuegra has reworked about half of his doctorate (defended in the University of Barcelona in March 1988) into a study of Russell's mathematico-logical career: the neo-Hegelian beginnings of the mid 1890s; the eventual acceptance of the set theory of Cantor and the studies of various algebraic theories by Whitehead (his former teacher and collaborator); then his acquaintance with the mathematical logic of Peano and his school; and finally Russell's first substantial outline of logicism in the book *The principles of mathematics* (1903).

A companion volume, drawing heavily upon the rest of Rodríguez-Consuegra's thesis, will describe the discovery of Russell's paradox. It will be more philosophical in character, and take that aspect of the story through to the 1910s, when *Principia mathematica* had been published and Wittgenstein was beginning to pontificate. Various papers on these themes have been published, and others are in preparation.

Dr. Rodríguez-Consuegra's work shows him to be an authoritative reader of Russell, both the published and unpublished writings: in fact, he has launched a veritable one-man Armada upon the history of Russell's logical thought. Unlike last time, the Elizabethans of today can only applaud such a distinguished contribution.

August 1990

Introduction

In this book I have reworked a substantial part of the materials of my doctoral thesis (*1987a*), which was devoted to the study of Russell's method throughout his whole work. In particular, I shall try here to show the origin of that general method in his writings devoted to the foundation of mathematics, by taking especially into account the writings of Peano and his school and the philosophical problems involved, which determined Russell's later evolution of.

I hold that this general method depended on the implicit belief that philosophy mainly consists in giving definitions. Starting from that, these definitions soon acquired the 'constructive' property (to take advantage of the connotations of this term, although with no relation with particular technical meanings established in logic or mathematics). Such *constructive definitions* always intended to offer an *analysis* of the concept involved into 'simples', presupposing the consequent *reduction*, although presenting it as a process respecting to some extent *ordinary language*, no matter the degree of *precision* obtained or the *loss of intuitiveness* produced. The implications of these five features and the difficulties related to the ontological 'elimination' of the entities defined, suppose a guide for Russell's philosophy, from his works on the foundations of mathematics to his metaphysics, through his logic, his epistemology and his philosophy of language.

There are mainly two reasons making necessary (and even urgent) a book like this. Firstly, there is no study of Russell's method starting from its origins in his mathematical philosophy, analysing and explaining its evolution. Thus, the existing global studies, either avoid the technical work of the first Russell (Eames, Ayer, Jager, Sainsbury), or do not succeed in isolating the methodological factors involved (Kilmister). In cases where more general methodological approaches are intended, either certain 'principles' more or less arbitrary and limited are introduced (Vuillemin), or the authentic unification of the method is not obtained, in spite of following the correct line (Weitz), just because the origins in the analysis of the logical and mathematical writings are eluded. This latter difficulty also affects other valuable historical studies (Kneale and Kneale, Bowne, de Lorenzo, Vuillemin, Moulines), especially because they consider Russell's doctrines mainly as a stage in an 'external' development, which makes it impossible to find the 'internal' appearance and evolution of the problems.

Secondly, the recent availability of the mass of unpublished manuscripts (*The Bertrand Russell Archives,* McMaster University, Hamilton, Ontario, Canada) makes it possible and demands an attempt to fill the gaps of the published writings, as well as to evaluate in a more definite way the actual influences and the underlying difficulties which were consolidating and developing Russell's methods. Such manuscripts (and correspondence) have been already used in order to throw some light on particular problems of Russell's work (Blackwell, Coffa, Garciadiego, Grattan-Guinness, Griffin, Iglesias, Spadoni), but rather with regard to very limited periods. In any case, the unpublished materials devoted to the foundations of mathematics before the Paris Congress of 1900 remained without a detailed study. However, only this study, together with that of Peano's logic, which I attempt here, can show the genuine roots of Russell's mathematical philosophy.

I come now to a brief account of the content of the book. Chapter 1 offers the context of the basic influences of the young Russell, which are usually neglected from a general viewpoint. In 1.1 I consider some traits of the logic of Boole and Peirce with the aim of understanding the context from where the logicist idea originated, although at the beginning only as the mere 'fusion' between logic and mathematics. This permits us to give a list of achievements which could easily be attributed to Frege, Whitehead or Peano.

Section 1.2 is devoted to Dedekind and Cantor. To Dedekind for having espoused views influential upon Russell's logicism, as well as for having proposed definitions (irrationals through 'cuts' between classes of rationals; the infinite through a relation of correspondence between a class and one of its subclasses) which exhibit the power of a certain 'constructive' method. To Cantor for being one of the fundamental precedents of Russell's particular type of logicism (as well as of Peano's), especially because he reduced arithmetical operations to set-theoretical ('logical' for Russell) relations between classes, and built transfinite numbers in a way according to which the infinite was introduced only after a series of important 'logical' definitions. Besides, some of his other achievements (the mapping between line and plane and the definition of continuity) show again the power of this kind of definition. Russell's difficulties in accepting Cantor's transfinites just till the moment when he started his logicist line are a good sign of the importance of this influence.

In 1.3 I consider another important antecedent of Russell's first attempts to found mathematics: Whitehead's 'universal algebra'. The most important thing here will be to point out the way how Whitehead already embraced a certain reduction of mathematics to logic, at least in so far as he shows how both sciences start from a very small set of common operations which allow their differentiation only in a later stage. Moreover, the development of Whitehead's work can be an instance of the 'transition' from abstract structures to more concrete ones by means of the mechanism of interpretation and its various applications. I also include a brief exposition of Couturat's ideas, which influenced Russell mainly through an incipient theory of logical construction and the statement of the concept of order as a logical relation previous to quantity (following Cantor).

The philosophical context is considered in 1.4, through Bradley and Moore. The first one for having provided the ground to Moore's relational theory of judgment, as well as for his attacks against the subject-predicate pattern and his method explicitly practised: to presuppose the existence of genuine 'logical forms' to be isolated for the philosopher by avoiding the bewitchment of ordinary language (although also respecting it). The second for having provided the majority of the relevant philosophical doctrines needed to make the 'liberation' from neo-

Hegelianism possible. That included, not only a new theory of judgment (adapted to realism), but also a vision of knowledge as an 'external relation', an atomistic (Platonic) ontology and a philosophical method based on definition (as analysis and reduction) starting from ordinary language. Of course all these recourses were immediately accepted by Russell.

Section 1.5 indicates a little of the main line of the first great work by Russell: FG; especially from the methodological point of view (for a detailed study see my *1990a*). Although in this work Russell's method continued to depend on Kant's philosophy, at least it shows some of the future traits, in spite of the strong criticisms against geometrical technical constructions.

Chapter 2 is devoted to the three main unpublished attempts to construct a philosophical foundation to mathematics, which were composed (before the contact with Peano took place) from some ideas already close to logicism. In them all the mentioned antecedents met in a common trend, although Moore's philosophy has the pre-eminence since the most important thing was to look for the 'indefinables' of mathematics as the raw material of the construction. The fact that numbers are regarded as indefinables and that Cantor is still rejected (although within a clear trend of progressive approach) was not an obstacle to exhibit an ordering of the themes already very similar to that of POM, and to regard relations as an indispensable recourse. On the other hand, relations gave rise to the first precedent of the 'principle of abstraction', which was later transformed into the true philosophical foundation of Russell's method of constructive definitions (some of whose basic traits are already present here).

The essentials of the method were already present before the contact with Peano. In particular, it is present the (Moorean) idea that philosophy is mainly an analytic activity that, starting from ordinary language, intends to discover, through immediate intuition, the fundamental terms of any field of our knowledge (the indefinable or primitive concepts) to present a chain of definitions constructing the rest of derivative concepts. This fusion of logic, ontology and epistemology never left Russell's project.

Chapters 3 and 4 explain in detail what the truly essential in Russell's study of Peano's works was, and what the achievements to be adopted were, according to his already formed general purposes. It has been, then, necessary to present an exposition of such achievements from the logicist viewpoint (chapter 3). The most important of them was the attempt to state a unique chain of definitions that, starting from a small number of primitive logical ideas, can present the main ideas of arithmetic, analysis and geometry as subsequent constructions out of those ideas, although of course introducing some more primitive ideas characteristic of these three parts of pure mathematics (cardinals, reals and vectors). However, the emphasis of Peano and his school on definitions, and especially the endeavours of some of the followers to reduce all kinds of definitions into only one (nominal explicit definitions), together with the application of this idea to the logical definition of the supposedly indefinables of mathematics, gave Russell the main ingredients he needed to complete his plan.

The writings after the Congress of 1900 began to develop this plan, until culminating in POM, a stage which I explain in chapter 4. Here the important thing has been the precise explanation of the readaptation of the former ideas to the new methods, by showing how POM (and to some extent also PM) was, first of all, a 'philosophical' version of Peano's *Formulaire,* including the same general structure (logic, arithmetic, analysis and geometry), although characterized by an explicit effort to dispense with the several gaps which made Peano's chain of definitions discontinuous. To do that it was sufficient to apply the method of transforming non-explicit definitions (by abstraction or by postulates) into nominal definitions, that is to say, to follow Peano, Pieri and Burali-Forti. That will be the moment to establish with precision what Russell learned from Peano and his school.

Finally, chapter 5 goes deep into the main philosophical elements of Russell's method, as exhibited especially in POM. This analysis shows its internal difficulties, which were finally unavoidable due in particular to the several tasks to be carried out by the principle of abstraction, i.e. the main philosophical ground of the constructive definitions offered. This principle was used, no matter how implicitly, to justify the replacement of entities by classes, as well as of symmetrical relations by asymmetrical ones. However, this constituted an excessive requirement of a principle limited to state such equivalences by means of the relation of membership not always well distinguished from other involved relations. Besides, the relational ontology Russell attempted in POM was often a difficulty for the needs of a logic based on propositional functions, which seemed to follow the subject-predicate pattern to some extent. Finally, this same ontology caused problems in geometry, where the tension between the pre-eminence of relations (required by logicism) and that of terms (required by the implicit atomism) showed a great difficulty to be taken on later.

As partial particular results of this research, certain unexpected philosophical and methodological debts have appeared. As a whole these debts show that Russell's dependence upon other authors is greater than the received view can concede. Besides, such debts belong precisely to the part of his philosophy that constitutes the true methodological unity, as it can be seen through the maintaining of certain single theses. To finish, here are some of these debts:

   - the initial general position against geometrical constructions (proceeding in part from Stallo and Hannequin);
   - the identification of a genuine philosophical method with the construction of definitions (proceeding from Moore), as a complement to the Bradleian search for the true 'logical forms';
   - the logicist idea (proceeding from Dedekind, Cantor, Whitehead, Peano, Pieri and Burali-Forti; Frege was known only later) as an application of a more general project: the construction of a chain of definitions starting from 'simples';
   - the method of transforming definitions by abstraction (and by postulates) into nominal ones (proceeding from Pieri and Burali-Forti), which gave rise to the celebrated definition of the number of a class in terms of a class of classes (which appeared, so it seems, before in Peano).

As for Russell's theories of descriptions and of logical types, they have already been considered in detail (also through the relevant unpublished manuscripts) in my *1989a, 1989b, 1990h* and *1990i,* and by other authors. These studies will be a part of a second book, which will be devoted to the philosophical applications of the method that is exposed here, and will consider other parts of Russell's work.

1. Methodological and logicist background

Here I shall try to reproduce, briefly, the main direct and indirect influences that served as a context for the origins of Russell's mathematical philosophy. Boole and Peirce (rather than De Morgan) constituted the main antecedent of all 'mathematical' considerations of logic, and then of all logicism. Couturat and Whitehead, whose influence on Russell could not have been more direct, added an essential element: the construction of all mathematics from the simplest operations and elements by trying to form chains of definitions (in the Peanesque style). With regard to Dedekind and Cantor, they were the main precedents of Russell's constructive methodology in the field of pure mathematical objects; besides, Dedekind explicitly agreed with logicism (without ever detailing his approach) and Cantor reduced arithmetic operations to relations between classes. On the other hand, Bradley and Moore created the philosophical framework providing, as methodological paradigms, respectively, the search for genuine logical forms and definitions as based on simples. In the last section I shall consider the basic ideas of FG, though only in so far as this work, previous to the influence of Moore, already shows some traces of the subsequent method.

1.1. Boole and Peirce

There is no doubt that the genesis of the different kinds of algebras made the birth of mathematical logic easier. Galois' theory of groups, by defining an authentic abstract structure based on properties and operations rather than on elements, was the remote antecedent to the later appearance of relations as genuine founding objects over the atomism of terms. The importance of the concept of *composition* also came to provide the necessary basis for obtaining a much more abstract viewpoint of classic arithmetical operations. Concerning this,[1] W.R. Hamilton with his

---

[1] Nevertheless, according to Bourbaki (*1969a*, 81) Galois' works did not exert an actual influence on the historical progress of abstraction in algebra, since they were rediscovered only in 1846, when already Grassmann and (almost) Hamilton had conceived their new theories. See a global presentation of the interaction between mathematics and logic from 1789 to 1914 in Grattan-Guinness *1988a*, and a 'philosophical' study of the matter in Vuillemin *1962a*.

quaternions and Grassmann with his calculus of extension broke the classical structure of algebra, and defined operations and concepts that, even though they were similar to those of arithmetic, enjoyed a greater generality by exhibiting less properties. Furthermore, W. Hamilton and De Morgan took the step from the intensional character of traditional logic of terms to the *extensional* viewpoint of the logic of classes, through the reduction of the former to the 'algebraic calculus' of the latter. With that we have the conditions that made it possible to conceive an authentic application of mathematics to logic, which was carried out by Boole through his 'mathematical analysis of logic' in 1847. Only in that moment was it possible to formulate explicitly and in its entire depth the problem of the relation between logic and mathematics.

For Boole the application of 'mathematical forms' to logic demands its consideration from a viewpoint outside of quantity, searching for 'a deeper system of relations' (*1847a*, 1). It will make logic stand out from the other sciences, showing itself as the only one capable of conserving at every moment a pure and formal character. The first famous paragraphs of the introduction make it clear that, in symbolic algebra, 'the validity of the processes of analysis does not depend upon the interpretation of the symbols which are employed, but solely upon the laws of their combination' (p. 3). For this reason, any interpretation is admissible as long as it does not modify those relations, and the same if it refers to arithmetic, geometry or mechanics. In this way Boole goes beyond the mere historical fact that the accepted 'forms' until then in mathematics were interpreted in a quantitative way, so arriving at a more universal view, i.e. 'to establish the Calculus of Logic, and that I claim for it a place among the acknowledged forms of Mathematical Analysis' (p. 4), even though, certainly, such a statement seems to place logic in the general field of mathematics.

This was nevertheless not the case: Boole's psychologism led him to consider logic as the pattern book of laws of the human mind, in so far as it limits itself to relate propositions and concepts in a formal way. It supposes that any *use* of formal reasoning involves a use of that machine of inferring that is our mind; thus, 'every process will represent deduction, every mathematical consequence will express a logical inference' (*1847a*, 6). With that at least mathematical processes of deduction depend on logic, although the mathematical analysis of logic supposed to present this latter science as a kind of applied mathematics. Besides, the logic with which Boole occupied himself was always an *equational* logic in which propositions have the form of identities starting from which relevant transformations are established; that is to say, a 'mathematical' logic in the standard sense of the term. That is why Boole does not decide on the question in a clear and definite way.

Boole notices that logic fails to belong to philosophy and has to be related to mathematics because, like mathematics, logic rests upon 'axiomatic truths, and its theorems [are] constructed upon that general doctrine of symbols, which constitutes the foundation of the recognised Analysis' (*1847a*, 13). But he also realizes that logic is different from mathematics since, as well as being a science, it also investigates its own foundations, which are merely presupposed in mathematics. Before the question of which one of them has to be regarded as the model of perfection, Boole does not decide: 'the exclusive claims of either must, I believe, be abandoned'

(p. 19). With that he states the *fusion* doctrine, a basis which made the appearance of logicism (a *reduction* doctrine) easier.

The following year Boole writes about the application of 'a new and peculiar form of Mathematics' to the operations of logical reasoning, which comes down specifically to the relations between classes and propositions, whose laws are capable of 'mathematical expression'.[1] In the definitive work, *The laws of thought* (1854) he insists that all operations in the language of reasoning form a system of symbols, operative signs and identity, which are ruled by laws being partially in agreement with the same ones of algebra (Jourdain *1910b*, I, 343). However, although he admits a similar structure, he claims that the isomorphism has to be established *a posteriori*, starting from some *independently* consolidated laws, on the ground that identity of laws is only an hypothesis. Nevertheless, he insists that the ultimate processes of logic are really mathematical, in a general sense according to which the essence of mathematics goes beyond number and quantity.[2] Thus, Boole avoids a definite decision about the possible pre-eminence of mathematics or logic.

The unpublished papers, as studied by Jourdain and Hesse, do not throw very much light on the problem. In one of them he writes about his logic expressed in 'mathematical forms', making it clear that this kind of expression is possible only in so far as we regard algebra, the science of number, as limited to *unity* and *nothing* (values with which the laws of algebra come true). That is why logic should be developed in an independent way, i.e. without falling 'under the dominion of Mathematical ideas' (Jourdain *1910b*, I, 346). With this aim, Boole had in mind to write an entire book about the philosophy of logic; however, from the published extracts (Hesse *1952a*) nothing new can be added to what we have seen in his work. Once again, the insistence on showing logic as being independent from mathematics appears,[3] even though this way seems to go deeply into psychologism (already present since *1847a*). This is confirmed mainly through Boole's belief in the possibility of establishing, by means of introspection, those general principles that, involving concepts, judgments and reasonings 'are anterior to and superior to all the special forms of its development' (quoted in Hesse *1952a*, 63). The interesting thing for us is to find, apart from many similar ideas, a passage (apparently intended for the introductory chapter to the projected work) in which Boole wonders, among other things, whether mathematical reasoning is peculiar or simply 'a special application of the ordinary rules of Logic' (*ibid.*, 64).

In any case, this question was doubtlessly answered in different ways by his readers and commentators, with which we can verify that the mentioned *fusion* between logic and mathematics was the phenomenon originally caused by the appearance of the 'algebra of logic'.

---

[1] In 'The Calculus of Logic', 1848; quoted in Jourdain *1910b*, I, 341.

[2] Jourdain *1910b*, I, 347. Already in 1851: 'Whatever, too, may be the weight of authority to the contrary, it is simply a fact that the ultimate laws of Logic —those alone upon which it is possible to construct a science of Logic— are mathematical in their form and expression, although not belonging to the mathematics of quantity' (Jourdain, *ibidem*).

[3] For example: 'to put the principles of the *Laws of Thought* into non-mathematical language' (Hesse *1952a*, 62).

The fact that one or the other of the two sciences was chosen as the genuine foundation depended simply on the way in which they were understood and, more importantly, defined. The one regarded as being wider and more general turned out to be the 'founding' science. If the fact that mathematics makes use of reasoning was emphasized, then logic was regarded as previous, but if the fact that those reasonings can only be expressed in all their generality using 'mathematical forms' was emphasized, then there was no doubt that mathematics was the basic science.

A sign of this twofold reading can be found by quoting different interpretations of Boole. For Harvey the term 'mathematical' was used by Boole 'in an enlarged sense as denoting the science of the laws and combinations of symbols, and in this way there is nothing unphilosophical in regarding Logic as a branch of Mathematics instead of regarding Mathematics as a branch of Logic'.[1] On the other hand, Bryant thought that Boole's fundamental point was having shown how logic underlies mathematics if we extract quantity from it. That is why 'Mathematics... is a particular kind of logic, and happens to be the kind which has developed to the fullest extent a language adapted for abstract thinking'.[2] Many of the later logicians can be placed on this ground, characterized by the original non-distinction and the twofold reading. In the following I shall refer mainly to Peirce, though I will mention Schröder and McColl as well.

Peirce was impressed by the abstract power of Boole's logic in relation to mathematics. However, even though he conceded an important role to logic in the foundation of mathematics, he never decided to take the logicist 'leap'; as we shall see later, that was surely why he distrusted the psychologistic danger implicit in any close relationship between mathematics and the 'laws of thought' (a danger to which, at least in part, Dedekind and Schröder succumbed). Already in *1867a* (3.20[3]) it is apparent that the common reason of his whole work on this theme is completely decided as we read that his objective was 'to show that there are certain general propositions from which the truths of mathematics follow syllogistically'. Nevertheless, such propositions, though mostly definitions of relations and logical operations (identity, logical addition and product), also include proper arithmetical operations (addition and multiplication) and numbers (zero and unity) even when these are presented as designating certain *entities* (*nothing* and *being*). Thus, the logicist idea is only partially present here. However, there are indirectly two more logicist elements: the first one is the interest in liberating the logical calculus from its dependence on mathematics (and so overcoming Boole's line), which will later be important for a clearer establishment of the relationship between logic and mathematics. The second element is the attempt to make use of the recourses of logic with the aim of obtaining more accurate definitions of mathematical concepts (both elements are mentioned in Barone *1966a*, 184, 211). These traits will remain throughout his work.

In *1881a* Peirce offers, on that line, something very similar to a logical definition of the concept of 'quantity', by transforming it into a *relation* with certain logical properties. However,

---

[1] Quoted in Jourdain *1910b*, I, 348; coming from a review of 1866.

[2] *Ibidem*; from a text of 1902. Such passages show that the logicist idea was present in England *before* Russell's published his important POM.

[3] I quote, in general, from the *Collected Papers*, by mentioning volume and paragraph.

even though the purpose of the work was also to show that propositions relative to number are 'strictly syllogistic consequences from a few primary propositions' (3.252), the problem of the logical origin of the latter is not established with precision (arguing that it 'would require a separate discussion').

An already clear connection between logic and mathematics appears in *1885a*. There (3.364) it is said that, as one of the objectives of the article was to enumerate the different kinds of necessary inference, the general procedure used for the algebra of logic will rest on the general categories of reasoning. For this reason Peirce adds that it is possible that his work could suppose 'a first step toward the resolution of one of the main problems of logic, that of producing a method for the discovery of methods in mathematics'. As a consequence, the character of *inferential* support of logic with regard to mathematics, already present in the rest of his work, is established. Nevertheless, as I stated above, precisely because logic must consider certain determining factors related to its *psychological* character, it does not offer, according to Peirce, sufficient guarantees of objectivity, or, as he preferred to say, a sufficient level of abstraction.

This was the position of *1896a*. There, starting from a Comtean scheme, it is stated that mathematics is 'the most abstract of all the sciences' (3.429), whereas logic, though more abstract than metaphysics (both forming part of philosophy) is still a 'positive' science, which, unlike mathematics, which tries only to preserve consistency, must admit in a compulsory way the existence of such things as doubt and falsehood. In this way, logic can be defined as 'the science of the laws of the stable establishment of beliefs', and *exact* logic as the doctrine of the conditions of that establishment, 'which rests upon perfectly undoubted observations and upon mathematical... thought' (3.429). For this reason logic depends on mathematics, and the idea that the former can supply methods to the latter becomes doubtful (3.454).

There are some other allusions to the same theme in *1897b*, where he returns to the idea of a logical theory of quantity, as well as in *1898a*, where he mentions the definition of Benjamin Peirce according to which mathematics is the science that *obtains* necessary conclusions (logic would be the science *of the obtaining itself*), and identifies this task with the handling of hypotheses (i.e. logical implications) (3.558-9). But although it is admitted that the difference between logic and mathematics is, ultimately, a matter of degree, it is again insisted that logic, as part of philosophy, has to take into account to some extent the reality of experience (3.560).

The final development and consolidation of Peirce's position, which, as we have seen, resembles logicism only in the sense that, like Boole, it does not state a clean, sharp distinction between logic and mathematics, takes place in *1902a*. This is an incomplete and unpublished (until 1933) work, as well as subsequent to the works by Whitehead and Russell whose framework of influences I am studying in this chapter, but it masterly summarizes the involved ideas. Peirce chooses, as usual, mathematics as a primary science, but following the same reasoning as Boole, i.e. starting from the fact that mathematics *uses* logic, although it also describes it and gives it its *forms* (4.228):

> It does not seem to me that mathematics depends in any way upon logic. It reasons, of course. But if the mathematician ever hesitates or errs in his reasoning, logic cannot come to his aid. He would be far more liable to commit similar as well as other errors there. On the contrary, I am persuaded that logic cannot possibly attain the solution of its problems without great use of mathematics. Indeed all formal logic is merely mathematics applied to logic.

With such a starting-point Peirce once again keeps his distance from the standard definitions in terms of quantity, and develops his father's celebrated definition of 1870 mentioned above.

The problem appears, of course, immediately: what kind of mathematics should we have in mind when trying to include logic itself within it? Peirce thinks that the character of necessity can only be reached in mathematics if we reduce it to the hypothetical trait common to all situations treated by mathematics: far from any *actual* relation to the world, 'Mathematics is the study of what is true of hypothetical states of things. That is its essence and definition... Conversely, too, every apodictic inference is, strictly speaking, mathematics' (4.233). However, this necessity, Peirce adds, does not come from mere deduction, which only draws *corollaries* out of certain axioms. It would rather consist of the one implicitly found in certain abstract schemata applicable to multiple situations (here Peirce seems to be thinking of 'structures'), and this calls for the construction of different 'theorematic' alternatives to be proved later. From that viewpoint, both mathematical abstraction and generalization have to be studied, in the same way as the connection between the two definitions mentioned above, where the first one will give us the method (to obtain necessary conclusions) and the second the objective (to study the truth in hypothetical situations).

Peirce explicitly rejects the pre-eminence of logic, even criticizing Dedekind (whose logicist tendencies, as we shall see later, seem to have been valued only by Schröder). He does it through a subtle distinction between the science of the *obtaining* of necessary conclusions (which would be logic) and the science that *obtains* such conclusions (which would be mathematics). With this he finds himself forced to describe logic as the study and analysis of the nature of reasoning itself (leaving for mathematics the establishment of the hypothetical propositions), turning it into a normative, categorical science in so far as 'it analyzes the problem of how, with given means, a required end is to be pursued' (4.240).

Now one sees the difference between both of them and especially the exact point where the formal connection justifying Peirce's view lies. For if logic wants to solve the problem of the relation between purpose and means, it has to employ mathematics, which means only that logic has a mathematical branch. However, this does not make any difference between logic and optics or economics, which also have a mathematical branch: 'Formal logic, however developed, is mathematics. Formal logic, however, is by no means the whole of logic, or even its principal part' (4.240). The reason is obvious: logic has to establish its aim and, in doing so, it immediately depends on the science of purposes (ethics), whereas mathematics never leaves the pure hypothetical framework.

There is no doubt that Peirce here talks in a context where 'logic' had a very different meaning that nowadays. One has only to compare the contents of the logical treatises at that time (for example, John Stuart Mill, Bradley) to see what they tried to embrace, especially the attempt to analyze and to go deeply into the 'nature' of judgment and of some relevant linguistic forms (subject, predicate, etc.). Peirce himself divided *exact* logic into three parts: speculative grammar, *logic* in the proper sense (which states the conditions of correspondence between assertions and reality), and *objective* logic (which would study the general conditions under which problems are presented and solutions are connected) (3.430). That is why Peirce emphasized this 'small' formal part as being the only mathematical one. In this way he assumes a clearly Boolean viewpoint which does not go beyond the framework where logic has a 'mathematical' wrapper when it becomes formal. However, also this fact seems to make clear the difficulty of reducing mathematics to logic: if logic is so wide as to embrace operations of a linguistic kind together with the basis of all reasoning (i.e. with the way these operate in our mind), then it depends to some extent on 'experience' (see above), making it impossible that a science so abstract as mathematics depends upon it.

To sum up: precisely because Peirce does not see clearly the possibility of a non-psychologistic logic, he does not accept the placing of mathematics above it, in spite of his constant recognition of the latter as *using* the procedures of the former. This does not mean that Peirce defended psychologism, in fact he defended the opposite view (Barone *1966a*, 202 ff); however, the danger of bringing mathematics close to everything related to the 'laws of thought' led him to avoid a fusion of doubtful consequences. This is the reason why a logicist vision seems to be prior for him, and at times precisely the opposite (the dependence of logic upon mathematics). In any case, the thesis of the *fusion* or, at least, of the extreme closeness remained, with him, secure, even though it was not in terms of the inclusion of one of them into the other (Thibaud *1975a*, 183). Finally, the connection making this fusion possible is the notion of *implication* as a basic characteristic of logic (and mathematics). The necessary conclusions referred to by Peirce are achieved through this relation between propositions, which is defined strictly in terms of the truth values of the involved propositions, i.e. in a virtually identical way as the modern implication (e.g. *1896a*, 3.439 ff; see also Thibaud *1975a*, 33-4).

Schröder's view coincided with Peirce's in starting from the recognition of a close relation between logic and mathematics. Nevertheless his psychologism, inherited from Sigwart and the German tradition in logic, led him to a logicism which, though recognizing the reduction of mathematics to logic, it depended on the mentioned 'laws of thought' (see Barone *1966a*, 198 ff). Already in the third volume of his major work on the algebra of logic he had attributed to Dedekind the merit of having filled the gap between logic and mathematics by means of his logical definition of number.[1] One might expect that in such a surprising logicist position there

---

[1] Quoted in Couturat *1900a*, 23. Curiously, Couturat attacks Schröder's view, though later he shares a similar Russellian position, and that with the same argument offered by Russell in the manuscripts: number is an indefinable intuition (p. 36).

had to be some influence from the 'pasigraphic' movement of Peano and his school, as one could infer from the following passage (from a paper of 1898 on this theme[1]):

> I consider pure Mathematics to be only one branch of general logic, the branch originating from the creation of Number... This view is confirmed by the fact, that under the pasigraphic aspect Arithmetic can do without any particular categories or primitive notions (such as multitude, number, finiteness, line, function, *Abbildung* or one-to-one correspondence, addition, etc.).

As Bowne pointed out (*1966a*, 18) the apparent contradiction between Schröder and Peirce did not seem to be noticed by these authors themselves in their mutual references. That probably was because they considered it more important to recognize that logic and mathematics belong to the same deductive system, than to establish the priority of one of them over the other.

McColl's logic is characterized by these same elements:[2] the non-distinction between logic and mathematics, the pre-eminence of the propositional interpretation (which must have directly influenced Russell; see Jourdain *1910b*, II, 225-7), and the use of implication as a fundamental relation (even over inclusion between classes). Like Peirce and Frege, he understood implication in an independent way from ordinary language, i.e. without any connection between the antecedent and the consequent, by meaning only 'if a is true, b must be true' ($a : b$) (Jourdain *1910b*, II, 221). As for McColl's defence of the pre-eminence of propositions in logic, it is historically important in spite of the fact that it depended on the need for using statements containing variables (see McColl *1905a* and *1905b*), i.e. what Russell called propositional functions (with some differences), to be able to give an account of the theory of probability, which was one of his main goals.

Finally, to illustrate McColl's view on the relation between logic and mathematics, here is a passage showing the way in which he saw the unity between them: 'Symbolic Logic (including Mathematics) may be defined as "the science of reasoning by the aid of representative symbols; these symbols being employed as *synonymous substitutes for longer expressions that are required frequently*"' (in Jourdain *1910b*, II, 235).

To summarize the logicist antecedents probably known to Whitehead and Russell before 1900, the following list of ideas already valid at that time is sufficient (excluding Grassmann, Frege, Dedekind and Cantor):

---

[1] 'On pasigraphy: its present state and the pasigraphic movement in Italy'. Quoted in Bowne *1966a*, 18.

[2] As we saw above these ideas were already frequent in England. Other evidence is supplied by Bôcher *1904a*, when he explains the definition of 'mathematics' by Kempe, coming from 1890 and 1894. The most emphasized thing in it was the importance which he gives to the notions of class and relation: 'If we have a certain class of objects and a certain class of relations, and if the only questions which we investigate are whether ordered groups of these objects do or do not satisfy the relations, the results of the investigation are called mathematics'.

- the idea of interpretation and the one concerning the diverse interpretations of an algebra or calculus;
- the parallelism between logic and mathematics, based on necessary formal reasoning which derives some propositions from others;
- the idea that a content can be expressed 'in terms' of formal language of mathematics or logic to take advantage of its deductive character (to which we must add the danger of the non-distinction between 'to express in terms of', 'to reduce to' and 'to deduce from', all of them present in Russell's later work);
- the pre-eminence of the propositional interpretation over that of classes, with the consequent overcoming of equational logic (though Whitehead applied it only to the algebra of logic);
- the pre-eminence of implication;
- the independence of mathematics with regard to quantity and number.

Here it is impossible to forget Russell's famous definition of 'mathematics' (in POM) as the class of all propositions of the form 'p implies q'.

1.2. Dedekind and Cantor

The concepts of *ideal, cut* and *chain* constitute Dedekind's great achievements in terms of the foundations of arithmetic, with regard, respectively, to the rigorous construction of algebraic, irrational and natural numbers. For this reason, they are also his main contribution to the methods for introducing concepts, which so greatly influenced analytic philosophy in general and Russell's work in particular. I shall say something about each one of these concepts, avoiding the details (see my *1987a*, 136 ff).

The *ideals* appeared due to the need for endowing rigour and a secure ontological ground to the *ideal numbers* which Kummer had postulated in order to solve the problem of the divisibility of algebraic numbers. In short, Kummer (*1847a*) introduced them as the necessary prime factors permitting the application of the fundamental theorem of arithmetic (the unique division into prime factors) to those numbers. On the other hand, Dedekind, avoiding this postulation, considered, not the ideal number, but the 'ideal' set formed by all numbers of the corresponding dominion that are divisible by a particular ideal number. In this way the notions of integer and divisibility resulted generalized to the new field, and that precisely by transforming divisibility into a logical relation, not between numbers, but between sets, i.e. starting the logicist line followed by Frege, Cantor, Peano and Russell.

As a process of *construction* it could not be clearer: it is carried out by making use only of simple, existing and known elements as raw material. Its objective is the definition of an 'object' having the required properties for the necessary fundamental laws (of arithmetic in this case), in a way that operatively 'exhibits' these properties as essential traits. Thus, it will be destined to replace other merely supposed or postulated entities.

The concept of *cut* appears in another context: the need for a rigorous foundation to the notion making it possible the existence of the limits being the ground of irrational numbers, in such a way that these numbers remain free of any recourse to intuitive representation. Dedekind carries it out by means of the usual standards for the 'creation' of other numbers: in the same way that fractions and negative numbers are to be reduced into positive integers and its laws, so irrationals are to be defined in terms of rationals, which of course requires knowing what is continuity (to fill in the 'gaps' in the series) through a precise definition.

It is well known that Dedekind achieves this definition starting from the analogy of the straight line divided into exclusive classes of points: there will then exist a point, and only one, that produces the division into two parts upon cutting the line. For irrationals, we shall consider any separation of the rational number system into two exclusive classes A_1 and A_2, which we shall call *cut*. Wherever we have one of these cuts: 'we create a new, an *irrational* number α, which we regard as completely defined by this cut (A_1, A_2): we shall say that the number α corresponds to this cut, or that it produces this cut' (Dedekind *1872a*, 15). After that we need only to put these numbers in order, through the relations between cuts, according to certain laws ruling the system, and reduce the operations with irrationals to others with rationals by defining the resulting cut.

In this way the notion of continuity remains rigorously defined and the irrationals constructively founded, despite the ontological ambiguities of Dedekind himself, who does not completely decide to *reduce* irrationals to rationals (hence the doubt between *to correspond* and *to produce* in the quoted passage; see my *1987a*, ch. 4). At least from the epistemological viewpoint, there is no doubt that the notion of cut acquires its strength through the previous existence of the elements serving as its foundation: the sets of rationals which make it possible. Besides, there is a clear parallelism to the *ideals*: the structural character through which two entities have precise properties, i.e. the ideal as a *system* and real numbers as a *system*, despite that in this way the limits of standard nominal and explicit definitions is overcome by entering into definitions 'by postulates'.

The same comes to light in Dedekind's construction of natural numbers, where the concept of *chain* appears. Starting from the three basic elements 'thing', 'system' and 'transformation' (mapping), Dedekind constructed an entire primitive set theory achieving once again his intended rigour. The fundamental definitions are the following (*1888a*). A *chain* is any system S containing an image of itself as a subset (for a given mapping of the system onto itself). The *chain of A*, A being a part (element or subset) of S, will be the common part of all those chains containing A. A system is *infinite* when there is a bijective mapping (a one-one correspondence) between S and one of its subsets, i.e. when S is similar (*ähnlich*) to one of its proper parts. A system N is *simply infinite* when there is a bijective mapping of N onto itself such that N is the chain of an element not contained in the image of N, i.e. if there is a mapping of N onto itself and a basic element such that: (i) the image of N is contained in N; (ii) N is the chain of the basic element; (iii) the basic element is not contained in the image of *N* (i.e. it is not the image of any other element of *N*); (iv) the mapping is bijective.

Finally, the *system of natural numbers* is a simply infinite system abstracting the nature of its elements and taking 1 as the basic element. With that (and after introducing complete induction, the definition by induction and the concept of number of a system) an abstract structure is defined: that of the class of simply infinite systems, which completes Dedekind's celebrated construction.

For the author himself, such a construction supposed a logicist reduction; that is why he already indicates in the *1888a* preface that arithmetic is a part of logic, especially when presenting the concept of number in a way completely foreign to space and time (although afterwards he falls into psychologism on presenting his laws as those of the human thought). This is what allows us to regard the new definition as a constructive one, since it searches for the simplest elements and, subsequently, reduces everything to them, showing that what had been regarded as simple is really complex under the analytic eye.

In fact, Dedekind's construction exhibits mainly two exemplary definitions: that of the infinite system and that of the chain of an element. The first one through the one-one correspondence with one of its parts, which permits him to manage the essence of the concept in a precise way, though breaking with the traditional intuitiveness and without drawing all the relevant consequences (that was the task reserved for Cantor). The second through the intersection of all the chains of which a given element is a part. This concept is also constructed on the ground of materials previously introduced and through an obvious breaking with the usual intuitiveness; first because it rests implicitly on the recourse to the infinite, and second because it is 'composed' by other concepts that can be hardly its 'parts' (they are 'greater' than the final concept to which they belong). One has only to remember Whitehead's method of extensive abstraction and the constructions that Russell and Carnap based upon it to realize the importance of this device.

The structural parallelism between the concepts of chain, cut and ideal has already been emphasized, mainly by insisting on the mathematical isomorphism among these constructions, which characterizes better than anything else the deep unity of Dedekind's methods (see Dugac *1976a*, 141-2; however, Dedekind himself did not point out this parallelism). For us, however, other traits are more important: the constructive problems usually resort to infinity; the introduced entities are limited to exhibiting the properties required for the role to be carried out; the logic and epistemological priority of the elements used as material of constructions; the problem of whether the constructed entities are to be identified merely with the laws ruling their working, and, on the other hand, whether we can apprehend them explicitly by their intrinsic nature; and finally the question of whether the constructions have to be mere scientific attempts to specify the ordinary language or whether it is necessary to resort to the creative capacity of the human mind (which would lead us to the problems of Platonism, eliminative reduction and the mere inference as the opposite to the genuine construction). In any case, all this takes us directly to a clear breaking with intuitiveness, which would be a typical feature of the later analytic philosophy.

We have something very similar with Cantor. Here I cannot enter into details, so I shall consider only some methodological implications of Cantor's contributions from the viewpoint of the constructive definition (i.e. the introduction of concepts in terms of ontologically simpler materials), independently of philosophical problems (psychologism, idealism, etc.). I shall consider irrationals, the cardinal-ordinal distinction, arithmetical operations, continuity, infinity and transfinites (according to Cantor *1883a*, *1895a* and secondary literature)

For Cantor, the usual consideration of irrationals as limits of infinite series of rationals fell into the logical mistake of *presupposing* the existence of these numbers through the idea of 'sum' of each one of these series. Such an equivalence, merely intuitive, should be eliminated by means of an alternative definition. Cantor made use of his concept of infinite *fundamental series* of rationals, each of which defined a number of the system of reals under certain conditions: each series had associated a *limit* that, though it was only a useful symbol destined to denote the property of making itself infinitely small in so far as the series increased, it would, however, be regarded as a 'number', even though it was only in order to filling in the irrational 'gaps' in the series of reals. In this way the problem of mathematical objectivity was solved, since these numbers did not have any meanings by themselves, except in so far as they belonged to existing objective sets. That is to say, irrationals received their objectivity from the task they carry out, despite the fact that in themselves they only represented the corresponding fundamental series. In this way the arithmetization of the analysis advanced one more step and the series of reals reached continuity. From then to their identification with the points of the line (see Dauben *1979a*, 40 ff), and the correspondence between reals and the points of the plane without resorting to intuition, there was only one more step.

The cardinal-ordinal distinction started out (in *1895a*) from the concepts of *set* (any group of distinct and well defined elements of our intuition forming a whole) and *power* (the concept derived from every set making abstraction both of the nature of its elements and of the order in which they are placed). From then on the concept of ordinal number (*Anzahl*) was introduced, also by abstraction: it is a special case of the wider concept of *order type,* i.e. the case of *well ordered* sets. Just as the cardinal proceeds from every well defined set, the ordinal does it from every well ordered set, i.e. by means of a single abstraction.

In this way the concept of *Anzahl* carries out two basic roles: to determine and delimit the concept of power and to make the essential difference explicit between finite and infinite sets, and this in a simultaneous way, since the concepts of power and ordinal coincide in the case of finite numbers (where the 'number' of elements does not change as the order changes) and diverge in the case of transfinites (where the same power can give place to many different ordinals). This complex relation between both concepts is also the result of the destruction of the traditional notion of number. Here Cantor (like Dedekind, Frege and Peano) realizes that the notion of number requires a logical foundation in terms of a construction, but in his particular view through the incorporation of the finite-infinite distinction: going up to the infinite, ordinals become separated from cardinals; going down from the infinite, they meet again in the finite (see Cavaillès *1938a*, 86). To sum up: the distinction between *Zahl* and *Anzahl* allowed the discovery of two different latent concepts under the notion of number, which led to specifying the finite-infinite distinction. At the same time, the distinction between various types of order served to clarify what can be truly called *continuity* and what cannot, thus making naturally possible the step towards the transfinites outside of the limitations of intuition.

We find the same foundational rigour in the reduction of arithmetical relations and operations into certain logical (set-theoretical) relations between sets. Cantor defined inequality between sets by means of the correspondence between one of them and a part of the other, through which it is possible to grant a precise meaning to the terms *greater* and *less*, and even to the concept itself of infinity. This is made only through the isolation of certain necessary and sufficient conditions which, instead of *explaining* the notion to be defined, draw its logical structure and its mathematical properties (e.g. uniqueness). The famous theorem B (Cantor *1895a*, §2) is a good example of this: 'If two aggregates M and N are such that M is equivalent to a part N_1 of N and N to a part M_1 of M, then M and N are equivalent', especially because the possibility so determined can only take place with transfinite numbers (and the subsequent and customary loss of intuitiveness of this kind of construction). From there Cantor achieved, on the same line, the reduction of addition, multiplication, exponentiation, etc., to similar relations between sets.

As for continuity, already in his *1883a* (§10) Cantor had offered a more or less adequate definition: the continuous set is *perfect* (all its points are limiting-points[1] and all its limiting-points belong to it) and *connected* (between any two elements there are an infinite number of others). In this way the explicit escape from the spatial and temporal intuition already appeared, searching once again for the necessary and sufficient conditions constructively allowing the specification of the concept through its logico-mathematical structure. However, in *1895a* (§10-11) a new definition was needed in so far as the former one still involved certain use of notions like *distance* or *segment,* which seem to belong rather to geometry. On the other hand, the definition of *1895a*, which cannot even be summed up here, had a completely general form, strictly in terms of *order* (that is why it has been classified as an 'ordinal' definition, opposite to the former one, which would be rather 'metrical'; see Couturat *1900c*).

The transition from finite to infinite suffered a similar evolution towards a greater logical rigour. In *1895a* the second number-class, like all the order types of well ordered sets of cardinality $\aleph_0$, is introduced, so giving rise to a constructed concept on the ground of pure order. The key to the definition lies in the recourse, appearing again and again, to the expression 'all the x such that...', which leads to the concept intensionally joining the elements. But I shall first briefly consider the more intuitive construction in *1883a*, then I shall compare it with *1895a*.[2]

The infinite series of the integers (v) is constituted, through the *first principle of generation,* always by adding a unit to the preceding number. In such a way a moment will arrive when, by taking into account the whole series in agreement with this law, it is impossible to refer to its greatest member. However, we can imagine a new number ω which expresses precisely the fact

---

[1] 'Limiting-point' (relative to a set) being that such that there are an infinite number of other points in any interval containing it; see Russell's clear explanation in POM, §273.

[2] In addition to *1883a*, I have used with profit: the letter to Dedekind from November 5, 1882 (Cavaillès *1962a*, 232 ff); Dauben *1979a*, 96 ff; Cavaillès *1938a*, 89 ff.

that the whole set is given (completed) through its law of formation (according to its natural order). Therefore we can regard ω as 'a *limit* to which the number v tends, if by that nothing else is understood than that ω is to be the *first* whole number which follows all the other numbers v' (*1883a*, §11). If now we apply once again the first principle to the new number ω, we obtain other numbers until we can imagine another number, which we can represent as 2ω and would be the first to follow v and $\omega + v$. Hence the appearances of ω and 2ω have been possible thanks to a *second principle of generation,* through which we can create a new number to be the limit of any determined succession of numbers in which there is no one being the greatest (*1883a*, §11). Applying now the two principles together we have:

$$3\omega,\ 3\omega +1,\ \ldots,\ 3\omega +v,\ \ldots,\ \mu\omega,\ \mu\omega +1,\ \ldots,\ \mu\omega +v,\ \ldots$$

until a new number is introduced (by means of the second principle) which immediately follows all the others, and that we can represent by ω^2. After that and following again the two mentioned principles, we would obtain new numbers, until we reach the immediately greater than all of them (through the second principle), which can be represented as ω^ω, and so on. Therefore, starting from the *first* class of numbers (*I*) (the class of the integers formed by mere addition), we obtain a set containing all the numbers preceding ω^ω, whose power will be that of (*I*).

On the other hand, the *second class of numbers* (*II*) will be all the numbers α we can reach by means of the two principles of generation:

$$\omega,\ \omega +1,\ \ldots,\ v_0\,\omega^\mu + v_0\,\omega^{\mu-1} + \ldots + v_{\mu-1}\,\omega + v_\mu,\ \ldots,\ \omega^\omega,\ \ldots,\ \alpha,\ \ldots$$

carrying out, besides, the condition that all those numbers preceding α, starting from 1, form a set of the same power as the class (*I*). For this reason, the class (*II*) has the power immediately higher to (*I*), i.e. the second power (*1883a*, §11). If we continue the same procedure we can reach a new number Ω, the first of a *third class* of numbers (*III*), and so on, up to the point of breaking every boundary in the formation of integers. The cuts in the series of transfinites are achieved by means of a *limiting principle,* according to which the successive numbers that are being formed, are all such that the class of numbers next to each class has a certain power: exactly the next higher power (as we have seen with the second class of numbers) (*1883a*, §11). The process of successively infinite generations is so limited until the different classes of numbers are obtained.

Thus, we have an extremely interesting constructive process in which two elements are to be emphasized, one positive and the other negative. The positive element is constituted by the precision with which *an entity is made to be equivalent* to the set of all numbers satisfying certain laws, or principles of generation, which remain formulated with great clarity. Thus, in an intensional way, all the elements of the series are obtained as a whole given once and for all by only carrying out the condition of remaining subsumed under certain law. Cantor rests mainly on this positive element when he states that 'the new number so obtained will then always be of utterly the same concrete determinateness and objective reality as the earlier ones' (*1883a*, §12).

The negative element is the effort demanded of the mind to 'imagine' new successive numbers that, by fulfilling certain conditions, come to satisfy the required properties. This is a clear recourse to the intuitiveness of a certain 'internal vision' (very likely of visual character) that would be capable of representing the *completed* infinite series. Although the numbers ω and Ω are clearly established as the first ones in their two respective classes of numbers, the lack of precision subsists as for the *application* of both the second principle of formation and the principle of limitation, which have to produce the necessary 'cuts' at the required point.

Like before with continuity, the difficulty is solved in *1895a* where ω represents an *order type*, that of the well ordered sets, from which the second class of numbers is introduced by generalizing all the order types of similar sets, in such a way that the principles of generation are shown to be a *consequence* of that order type (see Dauben *1979a*, 206). Thus, instead of resorting to the intuitive imagination, we now have a thoroughly abstract process, internally determined, which explains our capacity to represent the various classes of numbers. In this way, although some difficulties to *imagine* new numbers persist, the progress of abstraction has the virtue of *explaining* the traditional and intuitive formation of the introduced concepts through the *construction* of a previous structural model, instead of making use of intuition as a constructive instrument. These examples suffice to illustrate Cantor's constructive methods. As we shall see in 2.8 and 4.1, his influence on Russell was absolutely decisive.

1.3. Couturat and Whitehead

The relationship between Russell and Couturat began when the first received a letter from the latter concerning FG in 1897, and continued until Couturat's death in 1914. However, Russell carefully read Couturat's major work (*1896a*) probably already in 1896,[1] in order to prepare the review about it he published the next year (*1897b*). In the following I shall offer a brief summary of Couturat's work and then I shall select his essential ideas about definition and construction, many of which we shall find when we consider Russell's manuscripts in the next chapter. Finally, I shall refer to an article by Couturat which seems to be the methodological manifesto inspiring Russell's later work.

Couturat *1896a* is a work running between Kant and Cantor, characterized by an attempt of *philosophical* foundation of mathematics through the various generalizations of numbers as the main idea, until reaching the infinite number. By means of these successive generalizations, the author defends the general thesis according to which number and magnitude are independent (and

---

[1] According to Kenneth Blackwell's personal communication, Couturat's edition of *1896a* in Russell's library (now in the Russell Archives) is dated, by Russell's hand, in 'July 1896'. Therefore, the reading could have influenced even the definitive version of FG.

a priori) categories, and that those generalizations are an attempt to apply number to the different types of magnitude, until reaching the infinite. This notion, accepting Cantor's essential ideas (although rejecting for the moment the logicism implicit in the explanation of number in terms of sets), constitutes the supreme numerical generalization and the only one capable of providing an account of continuity without resorting to intuition.

The work is divided into two parts (each one composed of four books): the first part shows how the various generalizations of number take place in mathematics; the second part attempts a philosophical foundation of the process (on the Kantian ground of a priori necessity), culminating in a criticism of Kant's antinomies and a defence of the infinite[1] based on Cantor (for details see my *1987a*, 258 ff). I shall now allude explicitly to those concrete ideas which, being present in Couturat's work, played an important role in Russell's manuscripts of 1898-1900 as well as in his later writings.

The first of these ideas is the indecision between a formalist inspiration, a logicist trend and a certain empiricism. This latter appears when, as I pointed out above, Couturat assimilates the process of formation of cardinal numbers to that of the concepts in general, and it seems to be the reason (among others) that the general logicist trend does not be consolidated. Formalism can be found in a number of places throughout the work; for example in his description of the process through which the mathematician states the distinction between various types of numbers only by defining the operations ruling them. This makes Couturat say that the mathematician, unlike the philosopher, does not define magnitudes in themselves, but only their relations and operations (which are their mathematical or formal properties), until creating 'mathematical beings by means of arbitrary conventions; in the same way that different chessmen are defined by the conventions ruling their moves and relations' (*1896a*, 49).

This does not mean that Couturat accepts a merely 'formalist' construction of, for instance, arithmetic. He also demands the reduction to rigorous and precise properties to start from an intuitive basis, which is the same as conceding an 'intrinsic' significance to the ideas to be defined (*1896a*, 73). However, it is clear that it contains at least one element of arbitrariness. Another clearly formalist thesis takes place when Couturat, who claims the indefinable (and Platonic) character of the fundamental ideas of mathematics, writes that these ideas can be mathematically apprehended only by means of sets of axioms expressing the general properties that such ideas have to exhibit (p. 374). At this point the link between formalism and intuition appears: according to Couturat, to manage the reduction of an idea to a set of axioms it is necessary to *deduce* these

---

[1] In addition to Russell's review (*1897b*), which includes a good summary, Bowne *1966a* is limited to study the preface, the introduction and the final chapters where Couturat describes the discussion between the 'finitists' and the 'infinitists'. Besides, she presents Couturat as a convinced anti-Kantian, on the ground of his defence of Cantor, without realizing that Couturat continues to present certain categories as well as certain axioms as synthetic a priori —or 'rational'— judgments. That is why I hold that Couturat's main work is an hybrid between Kant and Cantor. The book recently devoted to Couturat (Couturat *et al. 1983a*) offers very little information about this work.

axioms from the corresponding 'rational' general idea that we all possess, which cannot be mathematically defined (p. 365).

Logicism is also present, although of course only as a still undeveloped trend; in fact there is no place in which Couturat states the possibility of reducing, for example, number to logical ideas (even in his *1900a* this possibility is still denied; see Dugac *1983a* and Dieudonné *1983a* for Couturat's later development). It is rather a logicism 'in blossom' which appears when Couturat writes about a pure 'universal mathematics', which would be the instrument of all sciences, describing it as a 'logic of quantity' (*1896a*, xxiii). Due to the possibility of such a science, Couturat constantly looks for a sharp separation between analysis and the idea of number, in order to arrive at the infinite without resorting to infinitesimal calculus, searching, like Cantor, for a purely logical foundation. However he does not dare to follow Cantor in his clear attempts of reducing number itself to logical notions (set, correspondence): it seemed to him (as we saw above) that this would imply going too far from the 'rational' pseudo-Kantian intuition which seems to grant the ontological status of number. Therefore, he defends himself from this logicist trend precisely on the ground of one of its traits: the conviction that there are certain primitive, irreducible and indefinable ideas (see Sanzo *1983a* for the philosophical significance of Couturat's logicism).

His position is somewhat clarified (until these indecisions disappear) if we resort to the genuine basis of Couturat's philosophy of mathematics (and of science in general): the theory of 'logical construction' (we even find literally this expression in *1896a*, 134). With that he provided Russell with an entire model to use for future works, as well as with a way of separating philosophical and mathematical definitions with clarity, even though the foundation he offered for this (the distinction between a rational order and a logical order) left much to be clarified.

Couturat already states a principle in the preface, taken from Cournot, which, according to him, is going to inspire his entire work; we shall call it the *Cournot-Couturat principle* and it says as such: 'the task of all Critique and of all Philosophy is to choose, among many chains of equally *logical* concepts, the most *rational*, i.e. that which introduces the greatest degree of unity, light and harmony in our notions, relating them to a few simple and primitive ideas' (*1896a*, x). The following example of its application is very close to what logical constructions meant for Russell some years later. Couturat starts from the distinction between verbal definition and definition of ideas: the first one is always arbitrary, therefore we can reject it, apart from its purely pragmatic value; as for the second, it tries to state an expression that we can compare to our previous idea. Thus, in his attempt to offer an acceptable characterization of integers as an alternative to empiricism (see above) Couturat says that we must not build the idea with concepts which do not imply it, but 'to describe this implicit idea under an explicit form, i.e. to offer, so to speak, a sign; in a word, to find a logical formula for this rational idea' (p. 333).

The proof that it is an application of the principle we are considering is that, to continue, Couturat interprets the example as an illustration of the 'critical and reductive method' necessary in philosophy, consisting of 'analyzing the fundamental ideas of science and going back to the general principles which are the starting-point of all deduction' (*1896a*, 334). The similarity to the idea of Russell's later logical constructions is that in both cases we have various alternatives from which to choose according to extralogical reasons. The basic difference is that while for Couturat the rational ideas are unique and irreducible, for Russell everything can be reduced to logical notions. Another difference, now comparing Couturat's method with Russell's after 1914, is that while for Couturat logical constructions do not have to replace in any way the rational ideas which they try to make explicit, for Russell they have precisely the intention of breaking away from the Platonism implicit in all 'inference' or 'postulation'. Curiously, Russell's manuscripts of 1898-1900 (see next chapter) took advantage only of that part of Couturat which was permitted by Moore's 'logic' (i.e. his views on concepts and judgment), which was his main guide at this time; that is why he was able to make use of this theory only much later, under the new influence of the ideas developed by Whitehead from 1913 onwards (see my *1987a*, 661 ff).

Paradoxically, it was in Couturat *1898b*, devoted to a critical account of Russell *1897c*, where all of these ideas are explained in a more articulated way, mainly by explicitly introducing the idea of *order*, which played such an essential role in Russell's manuscripts and in the rest of his work. Curiously, Couturat *1896a* hardly refers to order, although this idea was one of the most important for Cantor. In *1898b* Couturat overcomes this omission in the framework of a philosophy of mathematics[1] (similar to Moore's ideas) which since then inspired Russell: 'The first question to be raised upon studying the Philosophy of Mathematics is to know which are the primitive ideas that serve as material or foundation' (*1898b*, 436). The second point to emphasize is that, far from sheltering in the notion of number, he claims the need for adding to this set of primitive ideas also those of magnitude and order. However, as the idea of magnitude had already been developed (together with that of number) in *1896a*, he devotes the rest of the article to the idea of order.

We are not going to follow Couturat's reflections in this respect; it would suffice to say that, going back to Descartes, Fermat, Pascal, Leibniz and Bernoulli, he presents Galois as the genuine founder of the trend towards order through his theory of substitutions, of which he offers a summary (*1898b*, 437-8). Later on, by means of an attempt to relate this theory to algebra (which shows how order is independent of number and magnitude) and algebra to analysis (pp. 440-5), he arrives at the following main conclusion (pp. 445-6):

> If algebra is conceived to be the extension or the generalization of arithmetic, i.e. the pure science of number, we are led to this twofold conclusion: the algebraic resolution of equations resorts to the theory of substitutions; the numerical resolution of equations resorts to the theory of functions. In other words, the science of number cannot be completed or accomplished in any other way than through the science of order, on the one hand, and through the science of magnitude on the other. Hence the science of number is by no means self-sufficient; it is neither independent nor autonomous; it is

---

[1] It proceeds, according to Couturat's explanation in a note (p. 436), from his lectures about such a subject in the University of Caen in 1897-8.

compelled in its development to be lost and blended with other sciences, which, though doubtlessly are its neighbours, they are heterogeneous.

We do not need anything else; this is enough to fully understand from where Russell seems to have taken some of the ideas he used to compose the 'plans' of his various unpublished attempts to found mathematics before 1900. Besides, it happens that in the first attempt (AMR) the idea of order hardly plays an important role, whereas it is exactly since 1899 (FIAM), i.e. precisely when he knew about this article by Couturat, that his own foundational work suffered an important turn. Here the main trait was to give a fundamental weight to the notion of order, which, together with the previous ones of number and magnitude, came to state the ground to Russell's entire philosophy of mathematics until POM and PM. Besides, Couturat introduced the idea that the science of number is not enough for a philosophical foundation, which, although leads directly to logicism (as we have seen), paves the way for the doctrine that arithmetic has an important logical link with a deeper science: that of order. The only thing left to be noticed was that the science of order was no other that logic itself. This discovery was soon made by Russell, with the help of Peano and Cantor.

Nevertheless, the context of influences in the mathematical field would be incomplete without making a reference to Whitehead and, in particular, to his *1898a*. In the following I shall refer to his 'dominant idea' and, more briefly, to his algebraic structure, whose exposition is always accompanied by the one of the spatial interpretation.[1]

From our historical viewpoint, the preface to *1898a* is excellent from start to finish (all the following quotations come from this preface, unless otherwise indicated). It is an implicit proclamation of logicism and introduces, one by one, almost all the ideas with which we finished the section 1.1. As for the general aim and the idea of interpretation, Whitehead writes: 'It is the purpose of this work to present a thorough investigation of the various systems of Symbolic Reasoning allied to ordinary Algebra'. This will always go accompanied by the idea of a generalized conception of space, 'in the belief that the properties and operations involved in it can be made to form a uniform method of interpretation of the various algebras'. Thus, by investigating the possibilities of thought and reasoning, the comparison will be carried out through the unity of the only object of this *interpretation*: a general and abstract idea of space.

The definition of 'mathematics' starts from a vision of 'universal algebra' (see below for definition) according to which the latter is one of the branches of the former, as serious as the rest despite its novelty. This leads Whitehead, in the line of Peirce (and anticipating a similar view by Russell; see Grattan-Guinness *1990a*), to a vision of mathematics very close to logic: 'Mathematics in its widest signification is the development of all types of formal, necessary, deductive reasoning'. The pre-eminence of the propositional interpretation is as well confirmed when Whitehead adds that formal reasoning, which is the usual one throughout mathematics,

---

[1] There is very little literature on *this* work. I only know the following: Couturat *1900d*, Jørgensen *1931a*, Lowe *1941a*, Quine *1941a*, Bowne *1966a*, and González *1979a*, of which only the first one offers a complete survey.

completely dispenses with the meaning of the signs and confines itself to deduction between propositions: 'The sole concern of mathematics is the inference of proposition from proposition'. This is the reason why it has to limit itself 'to follow the rule', leaving the justification of this rule to philosophy (or to experience). In the second book Whitehead justifies this propositional pre-eminence, but without drawing the conclusion that we could have possibly expected: to concede the primacy to logic, and that with the argument that in logic addition and multiplication take place as well.

The idea of the independence of mathematics as to quantity and number is also explicitly present in the preface. For Whitehead, the fact itself of the possibility of a universal algebra is the result of a previous widening of mathematics, which took place with the appearance of complex numbers. Thus, mathematics was no longer confined to number, quantity and space, opening itself up to other deductive systems, based on entities originated by merely conventional definitions which fulfil the general, recognized laws. In this way the new algebras really represent *generalizations* of fundamental conceptions which, later, are presented as particular instances; therefore, Whitehead adds, such algebras are mathematical sciences which are not essentially related to number or quantity.

The *dominant idea*, i.e. the deep unity underlying the entire work, is based on the preserving of a constant spatial interpretation, in which the various developed abstract structures (algebras) are embodied. This makes it possible to look mainly for a global vision emphasizing the underlying foundation, rather than the exhibition of a survey exhausting all possibilities. That is why Whitehead points out that the intention of the work is to present the new algebras 'as being useful engines for the deduction of propositions; and in their several subordination to dominant ideas, as being representative symbolisms of fundamental conceptions... Thus unity of idea, rather than completeness, is the ideal of this book'.

Only this unity will make it possible to constitute genuine *new sciences* (since, according to Whitehead, every method creates its own applications) of the different subjects considered (from logic to the various geometries and mechanics). Therefore, it is a clear attempt to express in logical terms a wide part of mathematics by submitting it to a process similar to the one used by Boole, but making use of a machine as powerful as Grassmann's *Ausdehnungslehre* and by specifying the purely deductive-propositional nature of mathematics so regarded. This does not mean that the equational pattern is completely abandoned (one must remember that Whitehead was still depending on an obscure notion of *equivalence*[1]), but a unique structure (of which both logic and mathematics are parallel branches) is achieved *for the first time*.

The body of the work is composed of seven books in which the algebras and their corresponding interpretations are combined. Book I presents universal algebra properly said, and offers the basis for the two following: book II, devoted to the algebra of logic, and book III, entitled 'Positional manifold', which gives way to book IV ('Calculus of Extension'). The latter

---

[1] Quine (*1941a*, 130) has rightly criticized the notion of equivalence, still Bradleian, that Whitehead employs in this work, by calling it 'equivalence-in-diversity'. He adds, however, that this notion does not appear in his following works (perhaps due to the sound influence coming from Peano and Russell).

gives rise to a further division into other three: book V, where 'extensive manifolds' appear; book VI, which introduces a 'theory of metrics', and book VII, containing geometric applications. Let us then see first the development of the algebraic structure.

The first thing Whitehead does is to define the concept of 'calculus' in general terms: 'The art of manipulation of substitutive signs according to fixed rules, and of the deduction therefrom of true propositions' (*1898a*, §2); 'substitutive signs' being those which stand for something in the thought (§1). As soon as we have, Whitehead adds, a set of things related to some determined common property, we shall call it a *scheme*. When this later replaces others with which it shares equivalent features, we have *substitutive* schemes (§5). By assigning the properties of a scheme to a set of marks in the paper, we can use them as substitutive signs and fulfil in them the relevant operations. In this way, the mind will follow the rule without any need for using the imagination, and the calculus will be 'external' or demonstrative, i.e. without requiring the use of an authentic inference (§6).

Now *universal algebra* can be defined: 'is the name applied to that calculus which symbolizes general operations... which are called Addition and Multiplication'. According to the definitions we give of such operations, we shall obtain, either the simplest and most general algebra, or other *special* algebras (§12). General *addition* will be the unambiguous synthesis of the two things, producing a third one belonging to the same set, and following commutative and associative laws (§§14-15). Starting from it we obtain substraction and the null element (§§16-17). As for general *multiplication,* it will also be a synthesis, but its difference with the former is that it does not necessarily produce terms of the same algebraic scheme nor necessarily fulfils the referred laws, but the distributive one (§19).

This operation allows one to speak of different *orders* of algebraic schemes (Whitehead already uses the term 'manifold' here, which belongs rather to the interpretation). They will be of the first order when their elements can be multiplied together; of the second order when they are formed with the products of the elements of the first; of the third order when we obtain them through the products of the two first, and so on (always by means of the associative law). The set of all these schemes will be a *complete algebraic scheme*. When in an algebra only schemes of the first order can be produced, the system will be called of the first species (or *linear*). When in a special algebra some scheme of the mth order is identical to one of the first order, then we shall call it of the $m - 1$th species (since the $m + 1$th order would be identical to the second order, etc.). The 'calculus of extension' by Grassmann is, according to Whitehead, the only special algebra that can be of any species (§20).

By starting from this, the *classification* of special algebras is carried out (leaving aside ordinary algebra because it involves the notion of quantity, although its formulas are assumed: another step toward logicism) according to the following pattern (*1898a*, §22): *universal algebra* is the ground of all the following, and it is divided into two genera: (i) the non-numerical genus, which only includes algebra of symbolic logic, and (ii) the numerical genus, which includes two more species of algebras: the *linear* algebras (of the first order) and those of the *higher order*, represented only by Grassmann's calculus.

The basic distinction, i.e. that existing between numerical and non-numerical algebras, rests on two different meanings of the addition of a term to itself. In the numerical algebras we have $a + a = 2a$, whereas in the non-numerical ones we have a very different result: $a + a = a$. That is why symbols representing quantities are not used in the algebra of symbolic logic, the only one of this kind. The important thing for us here is to emphasize the form in which Whitehead has established a general kind of addition with the only aim of obtaining, later on, two other kinds giving rise, respectively, to logic and mathematics. With this there remains no doubt that some *fusion* between them is achieved through a common basis, even though this is done in a way that seems to attempt to overcome the approach used by former proponents of this thesis (Peirce, Schröder). This is the fundamental point to understand Whitehead's work as a precedent of logicism (see also 2.1 and 2.2).

As for the algebras of the numerical genus, they are obtained, as we saw, by adding the idea of quantity (including the possibility of imaginary quantities). All of these algebras have their general theory in the linear associative algebra (from B. Peirce) and they are characterized by the introduction of a new law for multiplication, that comes to be added to their usual general laws, producing so a much more general distributive law (in relation to addition). Multiplication would be the guide in dividing the algebras of the numerical genus into two species, in the same way as addition allowed two genera to be specified. Thus, a special kind of (combinatorial) multiplication will produce a special and unique algebra (Grassmann's calculus of extension). The rest of algebras (those of linear character) should have been considered in a second volume, including those by Hamilton (quaternions) and Cayley (matrices), which was never written. The fact that both Whitehead and Russell were preparing second volumes to their respective works (Russell was preparing the technical part to POM) contributed to the beginning of a collaboration that led both together to PM, once logicism was assumed as the philosophy to be explicitly developed.

Starting from this structure the various books of the universal algebra were developed. In the second book the *algebra of symbolic logic* is constituted, as a linear algebra, through the general laws of addition and the special law characterizing it ($a + a = a$). The next topic is the *calculus of extension* (bk. IV), since the theory of positional manifolds (bk. III) is already the *interpretation* needed to founding projective geometry on the laws of addition (in spite of the fact that it is based on a numerical algebra of the first order). As we saw above, it is a certain kind of multiplication which produces the new step: combinatorial multiplication; already within the numerical genus of algebras, i.e. assuming the idea of quantity (for details see my *1987a*, 248 f). The last three books are applications of the calculus of extension; for this reason they are interesting mainly to illustrate the interpretative and expressive capacity of this calculus according to the various 'manifolds', by giving an account of the several geometries and abstract mechanics.

There remains only the interpretation of the former algebraic structure, i.e. its significance. For Whitehead the investigation of the calculus must be made, if we want to progress, in connection with some interpretation, though he clarifies this Kantian position (which is based on intuition) by writing that the object of the investigation is algebra itself and not interpretation (§22). The basic concept from which all the interpretations are articulated is that of *manifold,* that is to say, the equivalent term to the one used by Riemann and Cantor, as Whitehead explicitly recognizes in introducing it (bk. I, ch. II). A manifold of elements is the same as what we before called a scheme of things, although now from the viewpoint of interpretation (still 'abstract') of that calculus; that is the concept which makes all the interpretative line of the work possible. This line is developed by means of the process of *particularization* of the most general manifold, according to the following plan: starting from general manifold (bk. I), one comes to positional manifold (bk. III), then to extensive manifold (bks. IV and V) and finally to spatial manifold and the several spaces (bk. VI).

The idea is that every one of them belongs to the former kind, but it is characterized by certain particular properties which are added to the general ones; the process is, therefore, identical to the previous one (which was purely algebraic). Thus there is a correspondence between the various algebras and geometries and the various manifolds, and this correspondence is strengthened by the fact that the idea of manifold is unambiguously *spatial* in itself. Whitehead adds that the series of interpretations will form an investigation related to the general theory of spatial ideas (established in bk. III) and, since even the algebra of logic will receive a spatial interpretation, we can see the work, to some extent, as 'a treatise on certain generalized ideas of space' (§22; see my *1987a*, 250 ff).

Up to this point we have seen the essentials of this fundamental work by Whitehead. As we shall see in chapter 2, the impact on Russell was so strong that it led him to the type of investigation and the kind of constructions which were one of the main recourses in his unpublished pre-Peanesque works.

1.4. Bradley and Moore

We shall now consider the authors who determined the strictly *philosophical* context of the young Russell, and who, on many points, coincided with the essential elements of the other lines of influence we have seen. In the following I shall point out mainly some methodological traits that may be found in Bradley and Moore (I attempt a more complete study of them in my *1990f*, *1990g* and *1990k*).

Philosophically, Bradley was the strongest influence on the first Russell. His antipsychologism gave Russell the necessary basis to attempt a foundation of mathematics far from classic British empiricism and, therefore, inclined to Platonism. His referentialist theory of meaning offered him a useful ground on the same line; however, this theory already carried the seed of future contradictions, though these contradictions came to be dangerous only after the abandonment of Hegelianism, in which all of them were assumed and 'overcome' in syntheses closer and closer to the Absolute. His theory of judgment, the bitter enemy of the subject-predicate pattern, put the basis for the atomistic theory of Moore and Russell, with only accepting the implicit arguments contrary to Aristotle and rejecting the ultimate reference of all concepts to a Reality from which they would be predicates. Finally, his usual practice of analyzing ordinary language, until finding the 'logical forms' underlying the mere 'grammatical forms', granted Moore and Russell a true method. This method needed only an atomistic ground to be transformed into that which was always the true philosophical method for both of them: the search for definitions showing the essence of concepts by reducing them to their simplest constituents.

I am convinced that these methodological elements are extremely important in understanding the rise of an analytic method in Moore and Russell. This method was always the way of arriving at *the kernel* of the questions and concepts, therefore it presupposed the existence of that kernel, either selecting the real form from the apparent (grammatical) one, or elucidating the diverse possible meanings of a term or judgment. This way of understanding the philosophical method brings us, finally, to a certain essentialism as intention and fundamental presupposition, since the standard questions like 'what is *x really*?' will be the usual ones. It is certainly difficult to elude this trend within the referentialist theory of meaning, since through this path one necessarily arrives at the identification between analysis and definition, just as at the consideration that penetrating into a concept means giving a definition of its true essence, as a unity composed of *simples*. With that, Moore and Russell carried on the traditional Plato-Aristotle-Locke line, but for Bradley the role of the simples remained obscure because of his relativism and his holistic and circular conception of language (which, incidentally, coincided much more with the present fashion than Russell's). That is why his analyses are not completely reductive like those by Moore and Russell. However, the usual elements of the Aristotelic pattern would be assimilated by Moore and Russell through Bradley's relational analysis of judgment.

There are mainly three Bradleian theses which can illustrate all this vision. They are quite difficult to be isolated, but may be clearly inferred from his philosophical practice. The first one says that there is a great difference between the apparent (grammatical) form and the real (logical[1]) form of concepts and judgments. It seems unnecessary to point out this doctrine in Moore and Russell. All of the early work by Moore seems to be based on it. As for Russell, his search for the *true meaning* was not limited, as we could think, to the stage previous to 'On denoting'; the partial breaking with Platonism that the 1905 theory of denotation involved served to accentuate the essentialistic tendencies: in fact they were not weakened until 1914 and did not *ever* completely disappear, despite the theoretical claims in favour of a linguistic conception of philosophy from the thirties onwards.

Bradley insisted in many places that the grammatical form is sometimes misleading, to the point of requiring an entire process to extract the true meaning from it. Thus when proper names play the role of true descriptions despite their apparent form; or when the subject-predicate pattern and the copula of the judgment are unmasked until being replaced by a completely different analysis. Likewise, Bradley was convinced that universal judgments are only grammatically

---

[1] Here I force the meaning of the expression 'logical form' a little, in so far as I attribute to Bradley certain connotations not actually present in his own expressions similar to this one. In doing so, I try only to emphasize a parallelism, but not to suggest a different Bradleian technical language.

categorical, while actually they are hypothetical; that existence is not a real predicate despite the linguistic appearances; and that judgments of identity are really tautologies (in my *1990g* I give more references and examples).

We have the same thing with certain concepts grammatically presented as corresponding to a true substantial (inferred) reality (like *substantives*) when, finally, they were nothing but linguistic forms whose only (constructed) reality was a function of simpler constituents (the supposed inferences relative to: self, soul, body, physical objects, etc.). These inferences are easily reduced to their ultimate constituents: those proceeding from the immediate presentation in our minds. This kind of analysis easily reminds us some Russell's recourses, like *acquaintance* and *egocentric particulars*; but the philosophical significance of 'construction' was very different for the two philosophers. For Russell (at least once he decided to break with certain essentialism) there are several acceptable ways of constructing the same concept: *we* as philosophers, explicitly carry out the construction. But for Bradley constructions are inferences unconsciously made; that is why the philosopher, with his analytic tools, has to avoid that they bewitch us as if they were substantial realities.

Let us now see some more examples that, either deepen the ones already mentioned or add more data. An especially significant one takes place when Bradley insists on his destructive analysis of the subject-predicate pattern, by denying that judgment is inclusion into a subject. He then adds (*1883a*, 22): 'By the subject I mean here not the ultimate subject, to which the whole ideal content is referred, but the subject which lies within that content, in other words the *grammatical* subject'. This doubtlessly supposes the *explicit* distinction between the apparent grammatical subject and the real or logical one. Thus, apart from the dubious nature of the 'real' subject, the important point for the method is the distinction in itself.

Likewise, by attributing 'true' meanings, which are different from the usual ones (that are only superficial), to other notions or judgments; by stating that 'now' really means position in time and, more accurately, 'simultaneous with' (*1883a*, 53); by reducing all terms of language into universals no matter how particular they seem to be (p. 63); by presenting analytic judgments (in a Bradleian sense) and singular judgments in general as truly hypothetical (p. 103-4); by giving the *true form* of the principles of identity (p. 143) and contradiction (p. 145), in both cases very far from their grammatical forms; and so on. We have, therefore, a method systematically applied and not only a number of isolated examples.

The second thesis is that only conceptual analysis can find out the real form from the apparent one and eliminate the latter. To show that this genuine foundation of analytic philosophy plays the same role in Bradley, I shall avoid any rhetoric by presenting only the textual evidence. In Bradley's terms, when we find ourselves confronted with a grammatical form that seems to go against a sound logic and, consequently, forces what we really want to say, we should resist the inclination of getting carried away by the linguistic appearances: 'in every proposition, an *analysis of the meaning* will find a reality of which something else is affirmed or denied' (*1883a*, 42; the emphasis is mine). Once this analysis is made, we can offer 'translations' different from the proposition that served as starting-point, which will be much more in agreement with that deep reality.

The same happens when, in trying to refer to something particular, we use the term 'this' under the belief that through this recourse we will reach the required referential unity. For Bradley this is an illusion: 'this' is also a universal; its only difference with regard to other universals lies in its incapacity of being used as a symbol in the judgment (*1883a*, 66):

> In every judgment, where we analyze the given, and where as the subject we place the term 'this', it is not an idea which is really the subject. In using 'this' we do *use* an idea, and that idea is and must be universal; but what we *mean*, and fail to express, is our reference to the object which is given as unique.

Of course only conceptual analysis can bring us to the awareness of such a mistake and to overcome it through the proper *translation* which puzzles out the real logical form.

In the case of singular judgments, which are really hypothetical, we are once again led by grammar to the belief that we are referring to individual objects. Especially when we use terms like 'this', 'that', 'now', etc. Bradley says that in believing so we make the mistake of thinking that we are dealing with a particular: 'But our real assertion, when we *come to analyze it*, never takes in the "that", or the "now", or the "this"' (*1883a*, 90; the emphasis is mine). What happens, and it is revealed by this analysis, is that we confuse two different ideal contents because of the misleading grammatical form. And we do it because the erroneous ideal content 'has not been analyzed'; when we carry out this analysis, 'the real judgment' and the referred object there appears at the same time (*ibidem*). It is clear enough, then, that if the correct analysis is not properly carried out, the grammatical form would lead us to gross mistakes *about reality itself.*

The third and last methodological thesis is that only through the analysis of the various meanings of a term can we know whether we are faced with something simple or constructed (i.e. capable of being eliminated). The thesis of the *true meaning* is not, consequently, only applicable to judgments. The need for finding the genuine logical forms also affects the true reality of concepts in themselves. In this second aspect of the same problem, the deep relationships between analysis, definition, reduction and capacity of being eliminated are shown, as they were inherited by Moore and Russell. Let us see some examples.

In logic, Bradley states that the true meaning of 'all' is not a set or list of particulars, but rather something hypothetical like 'any', 'whatever' or 'whenever'. In any case those terms are always found referring to some conditional 'if...' (*1883a*, 47-8). Therefore, 'all' can and has to be eliminated: it is something merely constructed and apparent. It does not contain anything categorical in its use; even though it is *reduced* to forms which are apparently categorical (*1883a*, 83).

In psychology, we especially have the splendid analysis of the self that is found in ch. IX of *1893a*. Through subtle and careful analyses all the involved meanings are shown, according to the diverse linguistic usages, until managing to build a true model to be applied, just as it is, by its inheritors (Moore and Russell). It starts from the need for going through, one by one, *all the possible meanings* of the term. The argument is already the standard one since (it is often believed) Moore, who, as it is well known, started most of his first articles in a similar way. Bradley says that the thing to do is to avoid answering questions about a supposed entity without fixing beforehand the meaning of the term that, we are told, designates it. The conclusion to the already famous investigation is that the general identity of the self is a senseless claim, since the question through which we might pose our problem would presuppose something already lacking any meaning (*1893a*, 73). That is to say, the term 'I' has no real meaning; consequently, not being a simple entity, it has to be discarded as an existent object (*1893a*, 81). It is as if, carrying out the project merely pointed out by Hume, Bradley sets himself up as the methodological father of all the later analytic school of philosophy. Thus, although his ultimate philosophical goals were very different (and even opposite) to those of Moore or Russell, he doubtlessly provided them with some procedures that would later be essential for both philosophers.[1]

In physics, it is worth mentioning the example of *nature* regarded as a construction (*1893a*, ch. XXII), especially since this analysis is useful in showing the need for *eliminating* any physical object as a merely apparent entity *in terms of its material constituents* (those supplied by the immediate presentation). The methodological starting-point is always the same for Bradley: what do we mean by x?, what does x really mean?, or what is the real form of this or that judgment? Obviously all that depends on Bradley's deep referentialism; thus, the notion of meaning remains transformed into the fundamental philosophical instrument, which has as a consequence that a definition has to be the only way of incorporating the result of the analysis. This essentialistic view was immediately accepted by the first analytic philosophy, which led it to its ultimate consequences (at least until the second Wittgenstein, who was never accepted by Russell).

The thesis of the construction and the consequent elimination, which leads us to the same general method is, however, twofold. On the one hand there is the possibility of analyzing inferred objects until they are shown as constructions out of 'simple' materials (ultimately proceeding from the immediate presentation). In this sense we can turn over the process and, by inverting it, explicitly achieving *constructions* making use of logical methods and applying them directly to the experience. This would be the type of constructive way followed by Russell from 1913 onwards. On the other hand there is the analytic view of definition according to which to carry out a definition is nothing but providing a conceptual analysis by emphasizing the constituents of a concept. Consequently, this would be possible only with complex notions, which can be *reduced* to their simple constituents. This is Moore's typical viewpoint; the only one available to Russell until he become aware of Peano and his discovery that this model of analysis had important flaws. We shall later see how Russell, through his constructive definitions from POM on (and his subsequent theory of descriptions), began to be aware that both senses are the same one.

Moore played the role of transforming Bradley's doctrines so as to make them usable within analytic-atomistic philosophy, and therefore for Russell's main purpose: offering a philosophical

---

[1] Since an identical analysis is applied subsequently to the concept of 'soul', I shall not insist on the need for eliminating this concept in a similar way, once it is shown as mere construction; see *1893a*, ch. XXIII.

foundation of mathematics which would show it as an absolutely true science. Antipsychologism and referentialism were accepted, including the rejection of 'ordinary' empiricism (John Stuart Mill). However, the conversion of the original idealism into a realism made it possible the adaptation to obtain that 'liberating' appearance which Moore and Russell always spoke about (although preserving the search for paradoxes), and at the same time, to avoid the dangers of an idealism that sometimes seemed inclined to Kant and, therefore, incapable of providing a strongly Platonic ontology.

As for the 'relational' theory of judgment, which characterized Moore's first philosophy, it may be summed up in one idea: all concepts of a proposition are to be regarded at the same logical and ontological level, together with the 'external' relations joining them, which must be seen as terms as real as the rest (*1899a*). This idea, inherited from Bradley with some modifications, allowed the rejection of the subject-predicate pattern and served to elaborate an atomistic ontology ruled by a single operation: the transition from what is simple to what is complex by means of the relation between the part and the whole. However, the logic seeming to be inferred out of the whole/part relation had at least two flaws: the non-distinction between membership and inclusion, and a kind of technical recourse which was hardly able to overcome Boole's viewpoint. These flaws made it impossible to successfully face the expression of mathematical propositions, which requires, among other things, the recourse of quantification, the rejection of equational presentation, and the reduction of syllogistic logic into a wider framework (which were achieved only through Peano; see ch. 3 below). The whole situation led Moore to the method of searching for the *indefinables* to reduce to them the rest of concepts through explicit definitions. I shall devote the rest of the section to this identification between analysis and definition.

The main ground and the first intuitions of Moore's methodology may also be found in his *1899a*, a genuine manifesto for all the subsequent logical atomism. The relational analysis made on Bradleian bases, together with the referentialistic theory of meaning, lead to a logic of externally related independent concepts. Ontologically they constitute the only reality, since they are the material of which the world is composed through the concept of existence. Our knowledge, as an external relation, conceives them (as well as their mutual relations) by means of the subject and the consciousness already at the same level as the other concepts. Intuition enables us to recognize the true relations between them; consequently, to understand something supposes and necessarily implies to know the concepts and relations composing it. Moore sums up his epistemology and ontology in this way: 'A thing becomes intelligible first when it is analysed into its constituent concepts' (*1899a*, 182).

With this formulation he already sets the ground for the construction of his definitive method, which always was (in some sense) analytic. Besides, this analysis, theoretically still somewhat limited and little used in practice, already supposed that conceptual complexity is the most important thing. We must not forget that in his ontology Moore made it clear that there is no separation between concepts and existents: existence itself is a concept. Therefore material complexity is derivative and secondary when compared to conceptual complexity (another doctrine that would entirely pass on to Russell). We have the same with epistemological complexity; its basic concepts (subject and consciousness) remain reduced to the same operative level as the rest. Here the simples are the constituents of judgment since this is nothing but a combination of concepts; even in case of an existential judgment. One must not lose sight of the fact that for Moore the object of any perception is an existential proposition; that is why there is no breaking between the given and the known (at least from the structural point of view, i.e. in terms of their relationships). All this set the basis for the logical, ontological and epistemological atomism that characterizes the beginnings of analytic philosophy.

In *1900b*, the first explicit theory on the relationship between analysis, definition and ordinary language appears (though it was partially present in *1900a*). Moore writes here, like later in *1903a* about the meaning of 'good',[1] that the thing to do is not to describe *which things* are necessary, but to know 'what that predicate is which attaches to them when they are so' (*1900b*, 289). Nevertheless, we are told, this does not have to lead us to what he calls a (mere) verbal definition, which would consist in enumerating the various meanings of the term according to its ordinary uses. Such an operation would doubtlessly be a correct definition: 'for the only test that a word is correctly defined is common usage' (p. 289), but it is neither a problem of meaning nor a factual question, but rather about 'what the predicate in question is' (p. 290).

Taking this into account, for Moore it is not a verbal problem, since we can be sure of the relationships between certain things (through a certain predicate) and the corresponding term, whereas 'we yet may be in doubt whether there is anything in common between these various predicates and, if so, what' (*1900b*, 290). With this, Moore does not avoid the Socratic-Platonic position nor his presupposed essentialism; he even suggests that only the inductive consideration can produce the conceptual (non-verbal) adequate definition. That is why he adds that the reference to verbal usages is indispensable : 'I must examine the cases in which things are said to be necessary, before I can discover what necessity is' (p. 291).

We have here a very useful early summary of the theory of definition regarded as something objective, in the sense that it has to be *distilled* from the various linguistic usages. The theory still is not very refined; the theme of 'true meaning' is not developed, nor is the one of the explicit equivalence between definition and analysis, but one can clearly see how ordinary language is the basic source of conceptual analysis. The conclusion of the article makes it clear that all the apparent meanings of 'necessity' are to be *reduced* to one: the logical meaning according to which what is necessary is what is logically *previous*. As this logical relationship is *the simplest possible*, we can use it as a criterion in defining other kinds of necessity (*1900b*, 300). However, precisely because it is simple we can only *point it out* without trying to delimit it with precision or exhaust all its possibilities: 'It needs, I think, only to be seen in any instance, in order to be recognised' (p. 302). Of course, this is like saying that it has to be conceived through intuition.

However, *Principia ethica* is the work where all these scattered ideas reveal their definitive status and remain related in a stable and defined way, which was essential given the elusive character of most ethical terms. It is then not surprising that Moore chose, as a starting-point, a

---

[1] Without forgetting that the basic doctrines of *Principia ethica* were already *in nuce* in some preceding lectures from 1898-1899: *The elements of ethics* (see Rosembaum *1969a*).

simple, basic and therefore 'indefinable' term to build the rest of the ethical concepts. It was a very similar attempt to Russell's in POM (both works were published in 1903); the main difference is that, while Russell was able to make use of Peano's techniques and the inspiration of an entire mathematical tradition (Cantor, Dedekind and the geometric constructions), he was not able to offer a convincing and complete theory of definition (see 5.3 below). Moore, on the other hand, did achieve this theory, though we must admit that the mass of problems that Russell had to solve was far greater. In any case, the view of definition as a reductive analysis in search for simple constituents passed on wholly to Russell.

Moore starts by posing the problem of the supposed definability of 'good' as the fundamental question of ethics, and to which everything else will be reduced (*1903a*, §5). Like in *1900b*, here it will be important to overcome what is merely verbal, since, although a term must be used in principle in its customary sense, what we have to study is the denoted object: 'My business is solely with that object or idea, which I hold, rightly or wrongly, that the word is generally used to stand for. What I want to discover is the nature of that object or idea' (§6). Of course, Moore adds that one cannot expect any possible definition of that object; as a simple notion, it is not capable of being described except by someone who immediately knows it. The non-verbal definitions are possible only 'when the object or notion in question is something complex', and they operate by *reducing* this object to 'simplest terms', which are indefinable and only immediately knowable (§7).

With such a view, Moore can be comfortably inserted into the Plato-Aristotle-Locke tradition, according to which everything being complex is definable in terms of simple concepts, whereas these latter concepts can only be intuitively known. Consequently, 'good' is indefinable only from the *reductive* viewpoint, but not in a lexical or conventional sense. In fact, when Moore emphasizes that, his purpose is always to take into account *objects*, not ideas; but since he refers to them as what we know only in an intuitive way, he is forced to take for granted that *this is its way of existing*, being dangerously close to psychologism. For this reason he writes that 'good' is indefinable in the sense that 'it is not composed of any parts, which we can substitute for it in our minds when we are thinking of it' (*1903a*, §8), with which he seems to suggest a certain kind of psychological incapacity of *our minds*. In any case, to be a concept incapable of being analyzed is to be (*1903a*, §10)

> one of those innumerable objects of thought which are themselves incapable of definition, because they are the ultimate terms by reference to which whatever *is* capable of definition must be defined. That there must be an indefinite number of such terms is obvious, on reflection; since we cannot define anything except by analysis ...

We come so to draw the ultimate consequences of the view already expressed in *1899a*; like then, Moore now insists that everything is *composed* of such simple terms (§10). For the same reason, understanding something can consist only in reducing it, through a definition, to its ultimate constituent parts, which will be 'simple' in a logical, ontological and epistemological way at the same time. Definition will always be analysis, and this analysis will always be a structural enumeration of simple terms (concepts) known by our intuition.

However, Moore denies that one could talk of a *faculty* called intuition through which one can prove certain propositions or contents of knowledge (*1903a*, preface). However he does it in a somewhat weak way; what he is really referring to is only the traditional 'intuitionistic' sense according to which the correction of a behaviour can be valued independently of its consequences. Therefore, when he says that he understands by intuition only what we cannot *prove*, he seems to place the limit so high at the beginning that subsequently he cannot surpass it. Since if we identify what is intuited with what is lacking of any proof, we shall also need some criterion to divide all things lacking of proof into intuitions and non-intuitions, which would require a second intuition, and so on. What happens is that Moore frequently employs terms similar to 'intuition' without explicitly mentioning it, and sometimes, when he does mention it, he forgets his own previous insistence that his particular form of intuitionism does not imply the admission to a special way of access to truth (or knowledge).

Here we have some examples of such uses in *1903a*. In §15 he concludes, after the analysis of the naturalistic fallacy, that there is an object of thought that is simple and incapable of being analyzed, with reference to which we have to define the rest of them, and that the important thing with it is to 'recognize it'. In §74, when he denies that ethical propositions can be reduced to some other existential ones or to the subject-predicate pattern, he writes that whatever exists and whatever the connection is between two existents, the problem of the good of what exists belongs to a completely different field: they are *two* different questions, and this fact is immediately 'perceived'. Finally, in §§86 and 90, when distinguishing between intuition and reasoning, Moore admits that intuition can supply a reason to hold that some proposition is true, and this is equivalent to recognizing a cognitive role for intuition, which would then be capable of carrying *information*.

I shall draw, to finish, some conclusions with regard to Russell. The main one is the close relationship between the most characteristic philosophical theses of both thinkers. In the same way as for Russell's philosophy, logic is the pattern to which ontology and epistemology have to be adapted. In his logic Moore rejects, with Bradley, the analysis of the proposition into subject and predicate, replacing it by a relational theory in which the proposition includes at least two externally related terms. This leads one, in ontology, to regard the world as divided into independent and different concepts (terms), and to reject the 'identity in difference' (i.e. the internal relations). Epistemology, according to that, tries to separate subject from object, presenting knowledge as an external relation incapable of altering its component elements.

The second conclusion is the division of concepts into simple and complex ones. The simple ones would be known by intuition, and the complex only by analyzing them into their simple constituents by means of a reductive definition. In this way, the only possible method in philosophy would be definition (analysis) coming from intuition. This partially coincides with Bradley, who also made an effort to obtain the real essence of concepts by overcoming false grammatical appearances, in spite of the fact that he did not accept any kind of genuine simple ontological atoms. For Moore, the simples are the atoms in logic as well as in ontology and epistemology: in the same way that proposition (as a complex concept) is divided into simple concepts, the world is divided into concepts, and knowledge is fragmented into a multiplicity of external relations.

According to Spadoni (*1977a*, 218) Russell's reasons for accepting Moore's first philosophy were that: (i) it was not empiricist; (ii) it was not psychologist; (iii) it allowed the explanation of the ontological status of mathematical entities; (iv) it stated the possibility of a priori truth (needed for the foundation of mathematics in Russell's mind); (v) it recognized the importance of relations. All this is true but Spadoni forgets the essential point: the atomism (pluralism) and the definitional method which made the construction of concepts possible. Besides, most of those points already came from Bradley; Moore's role was to transform Bradley's method (to extract the real form from the apparent one) into a kind of reductive, atomistic analysis through the introduction of 'simples' (which were unacceptable for Bradley's holism).

It was this theory of definition which allowed Russell to build mathematical entities (with Dedekind, Cantor and Peano) and later the entities of other fields of philosophy and science. He only had to add the idea of identifying Moore's simples to the primitive logical notions of Peano's logic (see ch. 4). Thus, we could sum up this process in an easy formula: while Moore showed Russell *what* to do in philosophy, i.e. building definitions in terms of simples, Peano showed him *how* to carry it out in practice through his logical techniques and his theory of definition, which coincided, at least in the essentials, with Moore's.

We can now understand in a better way the enormous importance that Russell gave to the influence of Moore (especially in the preface to POM and *1904a*); but one should notice that this influence was obviously insufficient as an instrument in solving the problems that Russell had to face then. This is illustrated by the fact that, except the work on Leibniz, the rest of his philosophical projects, in particular the several attempts of founding mathematics, did not progress satisfactorily until he got in touch with the methodological factor coming from Peano and his school (in the above specified sense). Only then it became possible to unify the trend proceeding from Bradley's 'intellectual constructions' with Moore's analysis through reductive definition, until arriving at actual constructive definitions, the true axis of Russell's method throughout his life.

1.5. *Foundations of geometry*

FG was published in 1897 as the final version of Russell's fellowship dissertation. I shall first summarize each of its parts (four chapters and an introduction) to briefly explain what the book is, and then I shall point out some methodological traits. In certain sense this section should precede the former one, for FG was written before Moore's influence, but it is also true that Russell's first work still belongs to the Bradleian context, which I considered in 1.4.

The *introduction* poses the main problem through two distinctions. The first one separates a priori knowledge from what is subjective, trying to formulate the former in a purely logical way and leaving the latter to psychology. The second distinction comes out in trying to articulate a true logical *test* (what kind of appearance would be impossible by denying a certain axiom?); as necessary propositions are hypothetical (Bradley), if what is necessary in knowledge is what is a priori, then it is required to provide the foundation granting the involved necessity. This can be done in two ways: either starting with the existence of geometry as a fact and, through an *analysis*, discovering the axioms from which it logically depends; or accepting its object (the space or, rather, the *form of externality* as the general concept) as a factual basis and, through a *deduction*, arriving at the principles (axioms) which make this branch of experience possible. As both results are the same, the axioms are a priori in a *double* sense. (However, these two processes can by no means be identified, despite Russell's claims, to Kant's 'similar' procedures; see my *1990a*).

Chapter 1 offers a brief history of 'metageometry' (a domain whose properties are common, or previous, to Euclidean and non-Euclidean geometries; see my *1990a* for details) following Klein's three stages. In the first one, he describes how the independence of the axiom of parallels was demonstrated. In the second, the relevant results of Gauss, Riemann and Helmholtz are critically exposed, as well as their application to space. In the third, he already enters into the field of projective geometry through the redefinition of the concept of *distance* (from Cayley and Klein) and the introduction of the Absolute (a figure making projective all metrical properties) as the ground of various geometries.

Chapter 2 critically reviews five philosophies of geometry, emphasizing what has to be respectively rejected or accepted. About *Kant* Russell recognizes that metageometry has proved the impossibility of still regarding the inference of the a priori and subjective character as valid by starting from what is apodictic of geometry (since other geometries are also apodictic). However, the converse inference is partially valid since, although it has to dispense with the distinction between analytic and synthetic judgments (in Bradley's sense), it makes it possible to regard everything presupposed in the possibility of experience as being a priori; and certainly a *form of externality* is found presupposed in it, although it is not necessarily the Euclidean space. Kant's first two 'metaphysical' arguments at least prove this, in spite of the fact that 'trascendental' deduction is partially invalidated and reconstructed by Russell. From *Riemann* Russell rejects what he regards as a wrong quantitative view of space, which adulterates his definition of 'manifold' that, even though valuable mathematically, is opposed to qualitative geometry. From *Helmholtz* he rejects the view that geometry depends on physics, on the ground that this position supposed a psychologistic attack against Kant. However Russell admits that geometry involves a reference to an 'abstract' and unextended matter which, although different from that of physics, would play the role of providing the terms for the spatial relations. The previous criticisms are repeated with regard to *Erdmann*, in so far as this author followed Riemann and Helmholtz, though Russell adds some particular criticisms of his geometrical axioms. Finally, he rejects *Lotze* by accusing him of technical errors. However, Russell makes use of Lotze's approach on the possibility of metageometry to fix his own ideas: non-Euclidean spaces are possible in the sense that they cannot be rejected by any a priori argument; therefore, while they fulfil the logical conditions required for any *form of externality*, the criterion to decide will be empirical.

Chapter 3, philosophically the most important, is a development of the *twofold process* described in the introduction to FG, which is first applied to projective geometry (section *A*) then to metrical geometry (section *B*). Section *A* starts by an analysis of the main traits and recourses with which projective geometry becomes an a priori science of all possible space; an analysis that will have to end up in the axioms presupposed in itself. Although such a science resorts to coordinates, this does not involve any use of spatial magnitudes; coordinates are nothing but conventional signs to denote certain points. Genuine projective coordinates are based on the quadrilateral construction (von Staudt), which permits to obtain a true descriptive definition of the anharmonic ratio independently of measurement. However, any attempt to define the point or the straight line has to be circular (principle of duality) and leads us to the contradictions of the relativity of space. Hence the need for defining the point in a projective way, starting from its relations with other three points (projective anharmonic ratio). This brings us to the general conclusion that two figures projectively related are qualitatively similar.

The former analysis leads Russell to the *three projective axioms*: (i) relativity of space; (ii) infinite divisibility (*point* as null extension and straight line and plane as sets of points); (iii) finiteness of dimensions (otherwise geometry would be impossible). Such axioms, which reduce space to purely qualitative properties and turn it only into an ordered set of mutually external positions,[1] belong as well to any *form of externality* since they characterize a priori any possible space. Therefore, they serve as elements to build this form as a concept independent of any particular intuition (Grassmann) and destined to formally supply the diversity of the material of our perceptive intuition. The section ends with the corresponding a priori *deduction* of the same axioms from this *form of externality* and its properties, completing then the circle.

Section B carries out the same twofold process, but by adding the idea of distance and, therefore, those of quantity, measurement and motion. For the same reason an empirical element appears and the metrical geometry splits into Euclidean and non-Euclidean. Nevertheless, since the three previous axioms are a priori necessary for any measurement, they are preserved here (with some minor changes). The axioms giving an account of the especially Euclidean traits have to be, consequently, empirically added. The twofold process is developed here in the following way. As for the *analysis*, relativity of position (homogeneity of space) from section *A* is transformed into *free mobility*. This is the guarantee that magnitude can be applied to spatial bodies preserving their figures and sizes through superpositions (congruence will be the term to be used for spatial equality). The denial of this axiom would lead us to the absurd conclusion of admitting absolute position and the action of space on things ('philosophical argument'), as well as to accept that, on varying the figures through motion, experience could not serve to determine, through measurement, such variations ('geometric argument').

---

[1] See Couturat *1898a*, 362. In general, the summary he offers has been quite useful to me.

Conversely, since free mobility involves relativity of position and this is nothing but pure formal externality, we can *deduce* from it homogeneity and free mobility (as a necessary property of any form of externality). After this circle, the first axiom becomes a priori in a double sense. By means of strictly similar reasonings, Russell states the axioms of dimensions and distance (two points define a unique spatial quantity: their distance), which also coincide with those from projective geometry and, therefore, are as well a priori in a double sense. All three are *presupposed* in the possibility of any spatial measurement in itself, and they are also *consequences* of the necessary properties of any form of externality. Consequently, the a priori element of *any* geometry will consist of these three axioms, which are common to Euclidean and non-Euclidean spaces. The rest of the axioms of Euclidean geometry (those of parallels, three dimensions and straight line —two straight lines do not embrace a space) will be, rather, empirical.

Chapter 4 (the last one) draws the 'philosophical consequences' of all the preceding through two problems. (1) Concerning the necessary character (for experience) of some form of externality, which has been constructed as a mere general a priori conception (like Riemann), Russell holds that it has also to be given as an intuition, although not as a subjective intuition (Kant), but rather as a mere mutual externality between things. This externality starts from the non-existence of particulars (Bradley) and rests on the need of all knowledge for a notion of 'identity in difference' (Bradley again) which implies time and that mentioned form of externality. However, this does not suppose deducting the perceptive world from mere categories, but pointing out (due to the constitution of the mind) that experience would be impossible unless the world possesses certain properties. (2) As for the contradictions that relativity of space gives rise (the infinite divisibility, interdefinability between point, straight line and plane, and relational order of space), one must admit they are unavoidable. But they can be attenuated by admitting a new construction: an 'abstract' *matter* that serves as subject to the diversity that space makes possible, providing the terms needed for spatial relations (unextended atoms) and making possible the distinction between empty space and spatial order (which Kant did not conceive).

Now I shall point out some involved methodological traits.[1] From this view the starting-point of this work is Bradley's 'analysis' and, implicitly, Kant's theory of construction. According to the first one, most of our philosophical concepts are 'intellectual constructions', i.e. mere abstractions; and, although some of them can be better founded than others (by resting more directly upon immediate presentation), in the last analysis all of them fail if they intend to overcome the framework of what is merely 'presupposed' in this of that branch of our knowledge. On the other hand, for Kant the constructions are only possible in mathematics and geometry: only these sciences can create concepts, although always by starting from *intuitions*. Philosophy cannot do it since it is forced to use *concepts* (i.e. a discursive apparatus). That is why philosophy can only offer non-constructive definitions, i.e. mere explanations that in the last analysis cannot go beyond the application of concepts to phenomena (as well as the analysis of concepts in themselves).

---

[1] I follow here a little part of what I already stated in my *1990a*.

Russell does not seem to take into account this Kantian doctrine (which is found, somewhat uncomfortably, at the end of his *1781a*, in B 741 ff). However, he realizes that the intuitive element in any abstract construction is necessary, and this led him to provide the form of externality, which was only a merely structural construction, with 'matter', by giving it not only an ontological but also an epistemological import. Nevertheless Russell does not manage, with this, to elaborate a minimum theory of constructive definition, though it is true that his further efforts on this line can be interpreted as an attempt to escape from the duality between apriorism and empiricism which underlies FG. In this way we could regard the introduction of a *matter* made out of unextended atoms within the purest 'empiricist apriorism' (similar to that of Erdmann). At the same time, the form of externality can be regarded as 'interpreted' by means of that *matter*, with which the former would be the logical structure to (what is presupposed by) the latter.

At bottom, all that could have been possible even within Kant's philosophy, since for him intuition is also constructive, and the entities and relations he employed are not mere data acquired in a passive manner; contrarily, subjectivity, which is the ultimate source of truth, also possesses its own creativity (see Bonfantini *1970a*, 373), just like the spontaneity of knowledge. But Russell also needed something that Kant could not provide: that his axioms were completely a priori and, nevertheless, without any synthetic element. Only in this way they could give an account of non-Euclidean geometries and, besides, be applied to the intuitively given space.

Through Hannequin *1895a* (from whom he took, among others, the idea of introducing atoms to solve geometrical contradictions) he could have learned that concepts that are built by means of definitions do not have to be regarded as real, but, rather, as mere recourses that, due to their properties, have to be valued mainly according to their fecundity. Thus, when Hannequin 'justified' the introduction of atoms, he pointed out their status as mere concepts that, just as the straight line or the infinite in mathematics, they are the result of a definition; therefore they already can enter into science through their capacity of making the deduction of a number of properties possible (Hannequin *1895a*, 15). But Russell still could not accept the existence of concepts which are created from a mere definition (i.e. constructed) under the only argument of fecundity. As it is shown in his criticisms of geometric constructions, or in those he made against Cantor (see 2.8 below), in such cases he resorts to his distinction between mathematical and philosophical definitions, but claiming that any entity has to be philosophically justified before being mathematically introduced.

Russell's systematic criticism against constructions in FG seems to proceed from three main sources: Bradley's view against any 'intellectual and abstract' construction; Kant's arguments against the possibility of any 'construction' in philosophy; and Stallo's criticism (in his *1882a*) against Riemann and any attempt of identifying conceptual possibility and reality. However, this criticism took place even before FG, as for instance when Russell described Hamilton's principle (the reduction of causation to the impossibility of creation or destruction of matter) as something 'regulative' and not constructive. The argument was that it presupposes the abstraction of motion with regard to moving matter, which supposes its application, not to reality, 'but to an intellectual and abstract construction of the real, resting on the distinction of substance and attribute' (*1895a*, 249). Thus, he is trying to make Bradley and Kant compatible, mainly as a means of strengthening his arguments against empiricism[1] and 'subjectivism'.

In the last analysis, Russell's rejection of any construction in FG[2] seems to rest on his rejection of particulars (another inheritance from Bradley), since they seem to be the material for any construction. That is why he needed to break away from monism before accepting Moore's theory of definition (or any other theory of construction), and that is why he opposed Riemann's and Helmholtz's attempts to regard space in a *serial* way (in the same analytic tradition of Descartes). However, for him any construction out of points involves a number of contradictions that can be solved only at a higher level (FG, 64). Because of this, all constructions are mistaken: they are made out of 'false' elements, and generally tend to construct what is (by its essence) incompatible with any reduction. Once again Bradley and Kant against Dedekind and Cantor: 'any continuum, I believe, in which the elements are not data, but intellectual constructions resulting from analysis, can be shown to have the same relational and yet not wholly relational character as belongs to space' (FG, 188). This is one of the contradictions that may be only metaphysically overcome.

Let us see now how the distinction between metrical and projective geometry was affected by the problem of constructions. Russell's acceptance of the philosophical importance of projective geometry, which took place under Whitehead's influence (as can be read in the preface to FG), was difficult, according to his previous view on this subject. The main argument to reject such a geometry (*1896c*, 100) was that metrical axioms would be valid even though magnitude was regarded as non-essential (according to the 'projective school'). He gave then four reasons: (i) measurement is the requirement to other sciences; (ii) the reduction of metrical to projective geometry is carried out through imaginary numbers, therefore it is a mere technical (and not philosophical) device; (iii) projective geometry presupposes spatial position, which implies measurement; (iv) projective geometry cannot avoid the use of the three metrical axioms. However, all of these arguments, except for the third one, are maintained in FG, even though this work, which was originally planned only for metrical geometry, was finally extended to embrace projective geometry.

Russell *says* to give only a technical significance to the distinction (FG, 9), mainly because of the lack of intuitiveness of projective geometry.[3] However, he seems to recognize that it has a certain connection to intuition as he believes that the form of externality is implicitly contained in it from the beginning through perceptive intuition. Only in this way can projective geometry manage

---

[1] In the quoted review (*1895a*) he shares Heymans' view against 'Mach and others, who maintain a more empirical position' (p. 249), but that was completely changed in 1914.

[2] Which are described as 'mere mathematical constructions', or 'purely conceptual (or intellectual)' constructions; see for example pp. 27, 105, 135, 189 and 147.

[3] This gives a parallel form to his criticism of the use of imaginary numbers, that depends (in FG) on the argument of *1896c*. For Russell, the use of imaginary numbers lacks 'philosophical significance' since they have no intuitive support (p. 43); therefore they are only a useful 'fiction' based on imagination (p. 45).

a conception wide enough to overcome particular 'spaces', and therefore give an account of any possible geometry. Here a usual idea of the subsequent Moorean stage seems to be present: what is primitive (simple and indefinable) has to be intuitively known. Under this argument his notion of a priori acquires a new importance, for saying that the elements of projective geometry are the *common* properties of any space (and even that, under the form of axioms, they are presupposed in any geometric reasoning) is admitting them as a *kind of indefinable to which everything can be reduced*. This is true especially when the task of discovering them must be carried out by struggling against the tricks of other presuppositions that, like those of a metrical character, 'are so rooted in all the very elements of Geometry, that the task of eliminating them demands a reconstruction of the whole geometrical edifice' (FG, 118). This seems to be a task for a genuine *analysis* (and the corresponding construction), whose methodological scope is admitted by Russell when he expounds the traits of the *third* period of metageometry (see above): 'It begins by reducing all so-called metrical notions —distance, angle, etc.— to projective forms, and obtains, from this reduction, a methodological unity and simplicity before impossible' (FG, 28). As I pointed out above, these arguments are an illustration of what can be called logicism 'in blossom', which will be stressed in the unpublished writings from 1898-1900 and developed after the work of Cantor and Peano began to exert influence.

This possibility of 'reconstructing' the entire edifice of geometry already attracted the young Russell. Thus, although Russell did not explicitly employ Boole's idea of interpretation, we can regard the 'step' from projective to metrical geometry as an instance of that idea (although perhaps only as a mere 'application'), since it supposes that by embracing any space, then 'every symbolic proposition is, according to the meaning given to the symbols, a proposition in whichever Geometry we choose' (FG, 9). This was the great task attempted by Grassmann, and with more general goals than those of mere geometry, as well as by Whitehead: to use a structure that, step by step, could be embodied with more and more particular concepts through the incorporation of new properties. In fact Whitehead *1898a* was published only some months after FG, and we know Russell read parts of the proofs.[1] Thus, if we remember that Whitehead's work was in print for almost two years (Whitehead *1898a*, 573), it is easy to recognize the possibility that a probable previous personal influence (towards a conception interested in discovering increasingly general and abstract structures) was confirmed through that reading. In order to be able to appreciate the true value of Whitehead's work, Russell still needed to be free of Hegel and Kant, to abandon certain 'metaphysical' themes (see Bonfantini *1970a*, 419), and to transform his incipient logicism into a true reconstruction, not only of geometry, but of all mathematics.

But FG, as a methodological attempt, became a failure precisely because of the great dead weight supposed by its inspiring philosophers (Bradley and Kant). We can specify so pernicious an influence by emphasizing three causes of the failure: (i) the need for presenting the reasonings (the transitions) in a circular way; (ii) the insufficient appreciation of the distinction between logic and epistemology; (iii) the denial of admitting (not only theoretically) the possibility of a true

---

[1] Whitehead acknowledges that in his preface, mainly referring to 'the parts connected to non-Euclidean geometry'.

formal science in which we can build abstract structures that were completely independent of our intuitive and epistemological capacities. Not even in POM would Russell manage to be free of such dead weights (the distinction between mathematical and philosophical definitions would be a sign; see 5.3 below), although it is undeniable that the mastery of Peano's new logic poses all these problems at another much more operative level.

To finish, I shall consider the distinction between mathematics and philosophy. The rejection of geometric constructions is an instance allowing us to confirm an essentialistic ground. However, a paradox appears: only mathematics has —for Russell— a 'firm ground' (FG, 177) and philosophy is characterized by speculation (p. 92); however, in so far as mathematics incorporates quantity, it limits any possibility of penetrating into reality. As he wrote about the notion of manifold (FG, 69):

> For mathematics, where quantity reigns supreme, Riemann's conception has proved itself abundantly fruitful; for philosophy, on the contrary, where quantity appears rather as a cloak to conceal the qualities it abstracts from, the conception seems to me more productive of error and confusion than of sound doctrine.

Hence Russell's attitude, also paradoxical, before projective geometry. According to him this science is logically prior to metrical geometry (and therefore philosophically more important), but as it is not able to rest on the 'firm ground' of quantity, it incorporates metrical geometry only in a fictitious way, and the supposed reduction does not go beyond the framework of what is merely 'technical' and lacking intuitive support (FG, 46). We then have two different criteria of the value of a conception. According to the first one, Russell emphasizes what is previous and a priori; but according to the second he holds an essentialism of qualities (his view of measurement forced him to believe that we can only measure what is qualitatively similar[1]): 'Hence a knowledge of the essential properties of space can never be obtained from judgments of quantity, which neglect these properties, while they yet presuppose them' (FG, 64). We find, once again, the inherited conception, common to Ward[2] and Hannequin (on the ground of Bradley and Kant), according to which science is limited to superficially describing the reality of the world, whose basic traits (continuity, infinity) slip through their fingers as being irreducible.

---

[1] That is why he ends the defence of mathematics from the philosophical attacks by Lotze in this way: 'I must, instead, rejoice that Mathematics has not been imposed upon by philosophy, but has developed freely an important self-consistent system, which deserves, for its subtle analysis into logical and factual elements, the gratitude of all who seek for a philosophy of space' (FG, 108).

[2] Ward did not encourage in vain (as a Professor and through his *1899a*) Russell's scientific curiosity on a Kantian ground (see Spadoni *1977a*, 55).

2. The unpublished mathematical philosophy:1898-1900

This chapter is divided into three main parts. The first one (2.1) completes the drawing of the context of Russell's influences with regard to his first attempts to philosophically founding mathematics. The second (2.2-2.7), being the most important, describes, within the mentioned context, the evolution of the three most important unpublished manuscripts. These unfinished works show how Russell had reached positions near POM from the *methodological* viewpoint, though still lacking an efficient logic for his intentions. The third (2.8) describes and analyzes Russell's evolution with regard to Cantor's ideas, going from a complete rejection to a partial acceptance, which prepared the enthusiastic assimilation that took place from 1901 on.

2.1. The genesis of the 1898-1900 manuscripts

In the preface to POM Russell explains his interest in the foundations of mathematics as a logical process starting from the progressive conviction that the study of a science, and in particular of physics, shows more and more the need for a direct analysis of the involved mathematical and logical problems. However, his famous reference to Moore's philosophy as the only one which made his dedication to these problems successful, did not then serve as recognition of something already announced in the preceding chapter: the decisive influence of Whitehead and Couturat in this transition from the problems of physics and mathematics to their logical grounds.

Russell's review of Couturat *1896a* denotes only a position still greatly influenced by Bradley's neo-Hegelianism. Thus, he poses the question of 'the relation of number to quantity' (*1897c*, 112) from a strictly philosophical viewpoint, far from the later recourse to geometry and logic. Russell rejects Couturat's position in so far as the French philosopher accepted Cournot's principle as a methodological guidance (see 1.3 above): reason should not decide among logical possibilities, but should find alternatives to the suggestions proceeding from mere understanding (p. 114). Of course, all of Couturat's arguments in favour of infinity (including Dedekind's axiom of continuity) are rejected on the basis that continuity is necessarily contradictory (p. 117), and that the process itself of formation of transfinite numbers involves a contradiction (p. 118). His general rejection from a more or less holistic position also prevents him from taking advantage of Couturat's notion of undefinability, as well as of his plan (still implicit in *1896a*) of a philosophical reconstruction of mathematics (p. 117). Russell even rests on the claim that Couturat's defence of infinity depends on 'some undue hypostatising of relations' (p. 119), although Russell himself later conceived the intention of building a general foundation of mathematics starting precisely from 'external' relations (according to Moore's influence).

On the other hand, Russell's reaction to Couturat *1898b* (which was a reply to Russell's more genuinely Hegelian article: *1897c*) is much more favourable. In fact it is contained in a reply to a previous article by Poincaré, with which one can better detect that Russell was already on the right path and was taking advantage of the first opportunity to describe his incipient method. It will suffice to point out here that Russell includes there: (i) his definitive solution to the problem of the distinction between mathematical and philosophical definitions; (ii) an attempt to present projective geometry as a formal calculus (announcing a future work devoted to this subject in p. 707); (iii) the explicit recognition of the importance of the notion of *order* for all mathematics (p. 703);[1] (iv) the acceptance, already present in the manuscripts in which he was working, that any philosophy of mathematics has to start from indefinables (pp. 700, 703 and 704).[2]

Two facts can explain this change: firstly there is Moore's influence, through which Russell already had a pluralistic philosophy available, which explicitly searched for the true indefinables and their relationships as the ground to any analysis. Secondly, Russell had already known a formal logicist geometry: that by Pieri (he possessed a copy of Pieri *1898a* since March 1898; see 3.5.1 below) which probably prepared Russell's quick acceptance of Peano's techniques.

I now come to another of my goals in this section: to show the impact of Whitehead *1898a* on Russell at that time. In FG Russell had tried to reduce metrical to projective geometry by constructing a global 'axiomatic' framework giving also an account of non-Euclidean as well as Euclidean geometries (the latter as a particular instance, by means of the generalization of the concept of space). He held besides that, since geometrical axioms are known a priori, the whole structure rested on a logical ground (for full details see my *1990a*).

Whitehead *1898a* added, to these *two* Russellian ideas (of which the first one also proceeded from Whitehead), two more lines of work (see 1.1 above): one going from general to particular algebras (including the reduction of logic and mathematics to a primitive algebra); and the other starting from the general concept of manifold and giving an account, by progressively adding certain properties, of an entire series of manifolds that can outline and reduce *all* geometries. With that, Russell's two ideas are proved to be quite superficial. If we also consider the pre-eminence of

---

[1] This recognition of the importance of order can be seen, in practice, with only comparing the AMR contents (of 1898) to that of FIAM (of 1899); it does not appear in the first one, whereas in the second an entire part (with several chapters) is devoted to it.

[2] The correspondence between Russell and Couturat partially shows this influence. Unfortunately I have not yet been able to see all of it (see the Bibliography). In any case Russell and Couturat met (for the first time) in November 1898 (from the 6th to the 8th) in Caen (see the chronology in Russell *1983a*, xxxiv). It can be easily imagined the enthusiasm of the young Russell about the form in which Couturat's mathematical philosophy (for instance that of Couturat *1898b*) came to philosophically coincide with Moore's pluralistic trends.

the propositional interpretation and the reduction of the whole structure to only two operations (general addition and multiplication along with their properties), the resulting model necessarily had to impress the young Russell as an authentic paradigm.

There are several textual pieces of evidence of the influence of Whitehead's work. One of them takes place when, in describing his readings after FG, he explains that he read many mathematical works, and that he was led to Grassmann through Whitehead's book, 'which greatly excited me' (MPD, 30). We can hardly obtain any other testimony so lively, although Russell later points out that then he was more interested in applied than in pure mathematics. (Although three pages later Russell recognizes that Whitehead had always been regarded in Cambridge rather as a specialist in applied mathematics.)

We find another relevant passage in POM when Russell refers to the constitution of an authentic universal algebra emphasizing the need for inductively studying the various species (rather than searching for the essential principles), and then he writes: 'The mathematical portion of this task has been admirably performed by Mr. Whitehead: the philosophical portion is attempted in the present work' (POM, 376-7). However, it is necessary to think, since POM was published in 1903, that Russell assigned himself this goal *as soon as he knew about the work* of Whitehead, i.e. as soon as he was convinced of the need for constructing a philosophical foundation for mathematics.

Curiously, Russell never admitted this explanation of his intentions at that time. In the most relevant passage he only refers to Moore and to his own evolution coming from the attempt to found physics: 'Gradually I found that most of what is philosophically important in the principles of dynamics belongs to the problems in logic and arithmetic. This opinion was encouraged by my adoption of Moore's views in philosophy'.[1] But after our examination of Whitehead's work, which also contained a *reduction* of certain principles of abstract physics (which Russell mentions) to universal algebra, we have to see it as one of the basic reasons that turned Russell into a researcher in the field of the foundations of mathematics.[2] To my knowledge, no one had emphasized and explained this influence of Whitehead *1898a* on Russell so far.[3]

Couturat himself probably played a catalyst role in this process of influence. It has to be supposed that, although his review of Whitehead's work was published in 1900, his opinion about

---

[1] Letter to Jourdain from 1910, quoted in Grattan-Guinness *1977a*, 132.

[2] In my *1990k* I explain something of what Russell carried out in his attempt of a logic for physics, but he was still depending on Bradley's neo-Hegelianism; that is why his results are not assimilable to his general constructive method. Curiously, some years later (once PM already finished in 1910), Russell again came to work on the logical foundation of physics, again because of a new wave of Whitehead's influence.

[3] This is doubtlessly due to the difficulty of obtaining and studying the great amount of existing manuscripts. It can be expected that from its publication (especially in vol. 2 of the *Collected Papers*, edited by N. Griffin and A. Lewis) this influence may be more deeply understood. The sole reference that I have read to some of them was in Grattan-Guinness *1985a*, which is a global review of the type of existing manuscripts, and in some of the works by Garciadiego (especially *1985b* and *1985c*).

it probably was formed with only the first reading. And this opinion supposed a clear logicist interpretation (doubtlessly influenced by his knowledge of Peano and his school):[1]

> From a more general and even more abstract point of view, the algebra of logic appears to be the foundation of mathematics, or the most elementary branch of mathematics, prior to the sciences of number, order and magnitude, on which all of them depend and are tributary: it is the mathematics of whole and part.

We shall see in this chapter how Russell, in his unpublished writings, literally followed this interpretation. In one of them (AMR, 1898) he even had come to put the concept of 'manifold', which plays such a great role in Whitehead's work, as the fundamental primitive one, and in the rest of them the doubtful point always was to know which concept had to be used as the logical starting-point. But this doubt was solved only through Peano's logic, once Russell had tried all possible combinations, always being inspired by the general idea that there are indefinable concepts and indemonstrable axioms with which to express, and from which to deduce, all mathematics. Only the writings by Peano led him to see that these indefinables and indemonstrables can be found in logic itself and not in the ideas of manifold, number, etc. At this point Moore's influence, which urged him to look for indefinables *in language*, could constitute a difficulty, since it forced him towards the analysis of *the language of mathematics*, but not towards that of logic.

On the other hand, Russell was still a Kantian, and he continued to be so for some time. When he came to believe that Cantor was one of the essential elements to be able to go on with his work on the foundations of mathematics on logical grounds, he was led to regard constructions as the only recourse to overcome the mere 'metaphysical' postulations. Besides, in the last analysis constructions are made out of 'elements', and these are either (again) constructions or simples. But Bradley did not admit simples; that is why the first step in the new philosophy had to be admitting and looking for them, which Russell learned from Moore; the second step was using certain particular methods to *construct* entities out of these simples, which he learned from Cantor and Peano, once he recognized the basic problem through Whitehead.

But Whitehead also helped Russell to pose the problem of choosing between *concepts* and *axioms* as the genuine philosophical ground of mathematics. In POM Russell claimed that all of it can be *expressed* (constructed) in terms of a few concepts, as well as *deduced* from a few axioms. These 'two' logicisms, which he never separated, are already present (and hardly differentiated) in Whitehead's work and Russell's manuscripts. Whitehead refers incidentally to this problem in

---

[1] Couturat *1900d*, 340-1. Couturat insists later that physics has to be included as well: 'Les Mathématiques appliquées ont pour domaine la nature entière, et englobant progressivement toutes les sciences physiques, à mesure que l'enchainement logique de leurs lois se dégage, et permet de les rattacher toutes à un petit nombre de principes expérimentaux; tandis que les Mathématiques pures, toutes idéales et formelles, apparaissent comme une méthode générale applicable à toutes les sciences concrètes, comme la science du raisonnement pur et abstrait, en un mot, comme la logique universelle' (p. 360).

several places, but the references to the first kind of logicism seems to predominate. Thus, for example, when he writes that the elements of the first species of algebraic manifolds *can be expressed* in terms of numerical coefficients to give rise to positional manifolds (*1898a*, §61); or when he says that the basic propositions of mathematics are *easily expressed* by means of Grassmann's calculus (ch. VI, bk. IV). However he also writes, in agreement with the second kind of logicism: 'The theorems of Projective Geometry extended to any number of dimensions can be deduced as necessary consequences of the definitions of a positional manifold' (§68). I think simply that Whitehead did not consider it necessary to carry out the distinction. A sign of this could be the fact that when he has to introduce the concept of 'distance' (in arriving at metrical geometry) he does it through *three axioms* (§197), which rule the actual meaning of this notion. The result was a certain non-distinction between the explicit definition of concepts and the status of them as being 'implicitly defined' by the relevant axioms.

In fact it came to be a usual practice of mathematicians to consider the operations simply as what exhibits certain properties. (We still explain relations by means of their logical 'properties'.) This situation was inherited by Russell in the 1898-1900 manuscripts (and even in POM), but Whitehead developed a 'logicist' system rather belonging to the 'structuralist' kind: he tried to define general structures from which other 'algebras' can be presented as particular instances, whereas Russell, under the influence of Moore and his essentialistic kind of analysis, chose a very different approach. For him the most important thing was the search for simple (indefinable) concepts, but without renouncing to *explain* them and without admitting for them a status simply characterized by a series of postulates implicitly defining them. The reason was that he thought, like Moore, that simples are immediately known to us by intuition, and that the object of the investigation must also be the labyrinth of ordinary language, in spite of having only two main tools: Bradley's old referentialism and a still babbling theory of external relations, to build an ontological foundation of 'simples' (see 5.4 below).[1]

The root of this twofold tradition might be Grassmann's obscure view (from which Whitehead built his *1898a*). He clearly admitted the possibility of implicit definitions and he clearly distinguished between formal and empirical sciences (see my *1987a*, 71 ff); nevertheless he put the borderline between the two kinds of sciences in a way that geometry remained at the same side of empirical science since, according to him, it refers to something *given* in nature: space (see Nagel *1939a*, 169). Thus, as the formal sciences remained separated from all actual existence, they did not need to come from *axioms* (or 'truths'), but they should start from the implicit properties of conventional *definitions* (as Whitehead wrote in the preface to *1898a*), and develop these properties through the calculus. In the following passage by Grassmann there seems to be the kernel of Whitehead's mentalist philosophy (quoted in Nagel *1939a*, 169):

> Proofs in formal sciences do not go outside the domain of thought into some other domain, but remain completely within the field of combinations of different acts of

---

[1] Russell still continued in POM looking for these concepts; for this reason he rejected Peano's definitions 'by postulates' (from Gergonne and Pasch).

thought. Consequently, the formal sciences must not take their point of departure from *axioms*, as do the real sciences, but will take *definitions* instead as their foundation.

With all of this inherited set of non-distinctions and paradigms Russell rushed into the attempt to *analyze* mathematics to reduce it to its essential elements.

Once we have described the essentials of the genesis of Russell's attempts of philosophically founding mathematics, we now come to the content itself of the main relevant unpublished manuscripts. They are three: *An analysis of mathematical reasoning* (AMR, 1898); *The fundamental ideas and axioms of mathematics* (FIAM, 1899); *Principles of mathematics* (POM1, 1899-1900).[1] My fundamental aim will be to show how we can find in them the most important ingredients of what, from 1903 to 1948, would be the general philosophical method of Russell. As we shall see, in order to show the appearance of such ingredients in the framework of all mentioned influences the content of these manuscripts is vital to understanding the origins of such a fruitful and influential methodology.

However, with these unpublished manuscripts we have a similar situation as with FG. Until now, all of Russell's scholars, no matter how critical and subtle they have been, have avoided delving into these vast, prolix and somewhat tedious (unfinished) writings, by probably following Russell's own later judgment: 'I still have the MS of what I wrote on this subject just before my visit to Paris and I find, on rereading it, that it does not make even a beginning of solving the problems which arithmetic presents to logic' (MPD, 51). I am not sure that his opinion is justified. In any case, the detailed study of these manuscripts is indispensable to any attempt to reconstruct the genesis of Russell's methodological constant recourses.

2.2. Logic, mathematics and ontology

In this section I shall try to relate Russell unpublished ideas to the seeds of logicism we have seen in the previous section (and the former chapter). For this, I shall start by studying the evolution of his thought, throughout the manuscripts, on the philosophical *priority* between logic and mathematics.

In AMR the problem seems to start (in the introduction) in a very favourable way to a logicist position, since the content of the work is justified by resorting to the following series of 'transitions': firstly, we have the extension of concepts, i.e. the logical calculus, which is followed,

---

[1] For more details on the manuscripts, see the Bibliography. Feinberg *1967a* included a general description of all the manuscripts known at that time. As regards the form in which I quote them, I often quote the part and the chapter of the reference (the chapters are very short) instead of the page, whose number is at times difficult to be determined (for example: V/2 will mean part V, chapter 2).

through the introduction of the cardinal number, by arithmetic; later, we have the ordinal number and (pure and extensive) quantity, until arriving at geometry and dynamics. In the list of basic ideas of mathematics that follows the former plan, we find the same order: firstly, we have the concepts of *manifold* and addition, which are used in all mathematics; later we have number, order, relations of equality and inequality, dimensions and the idea of thing. In this way Russell seems to choose a plan similar to that of Whitehead *1898a*, with the advantage that, on placing the logical calculus at the beginning, he seems to regard it as logically previous to all mathematics. However, on defining 'manifold' as 'a collection of terms having that kind of unity and relation which is found associated with a common predicate' (I/3), he practically identifies it with the notion of (intensional) class. This is why he then introduces the concept of 'assemblage', which lacks that common predicate and can give way to an *addition* of mere 'terms', whereas classes would require a *synthesis* of predicates. In this way the object of logic is now the relation between these two operations: addition and synthesis;[1] and the notion of manifold remains to be less general than what it seemed to be in the beginning.

On posing the problem at this point, certain difficulties arise and the apparent 'logicism' does not seem to go beyond Whitehead's position. Russell regards addition and synthesis as mere particular instances of a general mathematical operation: traditional *addition*, i.e. a combination of several terms producing another (I/4). This agrees, besides, with his description of the logical calculus (I/5) as 'a branch of mathematics' and as the science of manifolds, i.e. as something 'less general' than arithmetic. The argument, clearly inspired by Whitehead *1898a*, is that numbers are not only applicable to manifolds (or classes), but also to assemblages, which are more general entities. Finally, he recognizes the existence of a twofold category, that of whole/part, capable of giving an account of all logic (the 'logical calculus'): calculus is that branch of mathematics that depends on the categories of whole and part. This has two basic consequences, both on the line of the fusion between logic and mathematics (see 1.1 above): the first one is the liberation of logic from the traditional predication, which would come (like in Whitehead) to be the result of a mere interpretation; the second is the recognition that the category of whole/part, in showing its fruitfulness in logic as well, would be extended to all mathematics.

On arriving at 'number' (AMR, bk. II), Russell writes, the generality of the logical calculus disappears, in so far as we go beyond the necessary relations among numbers and we apply them to what exists.[2] However, if we consider the judgments expressing 'logical' necessity it is

---

[1] As we saw in 2.3, Whitehead *1898a* avoided this problem because, by starting from a merely algebraic point of view, he came to logic after having already defined addition, and he regarded the 'predicates' (like propositions) only as a mere application of the calculus. On the other hand, Russell, as he wanted *to start from logic*, he had to consider from the beginning all the problems of the distinction of terms and 'meanings' for applying later the results to all mathematics.

[2] This application would form the 'existential judgments of number'. Russell's rejection as regards *necessity* is parallel to Moore's with regard to the 'existential theory of judgment' in general. For him, like for Russell in PL (see my *1990k*) judgments are only maintained if they are pure contents that show the relations of logical terms among themselves.

possible then to interpret, for example, addition in terms of these pure concepts which have to guarantee its logical ground (II/2). At this point Russell seems to return in full to what we regarded before as a logicist attempt, since in seeing numbers as 'logical subjects' and analyzing addition through the framework of the whole/part relation, he necessarily interprets numerical unit (number 1) as a *part* of a whole. That is why the addition demands the analysis into units, and addition is always addition of *terms* and not of numbers; and, although both operations could correspond to each other, the truth is that arithmetic deals only, for example, with 'two' and 'three', but not with the more general concepts of *duplicity* and *triplicity* of the logical calculus. With this amazing precedent of the later logicist definition of number, Russell is forced to draw the conclusion that 'the idea of a sum of terms forming a new single term —the idea of the extension of a concept, in fact— would appear to be logically prior to the necessary judgments of Arithmetic' (II/2). Therefore, the concept of manifold and the logical calculus itself would be *presupposed* in arithmetic.

The conclusion to AMR seems to be, then, that though mathematics is more general than logic (since the former proceeds from the general concept of addition), however an analysis of the latter quickly leads us to verify that the whole/part relation underlies it, since this is the only notion that can give an account of the plurality of the logical subjects that explains addition. Since the whole/part relation underlies as well logic itself, it must be concluded (in a similar way as Whitehead) that logic and mathematics have a *common basis*. However, when we analize this common basis, in so far as it is 'logical', we see that, though it is prior to arithmetical relations among numbers, it is not prior to the numbers themselves, which are logical subjects too (Russell writes 'pure meanings'). We must drop the problem just in this uncertainty, which is ontological in the last analysis, for it depends on the relations between numbers and mere 'terms'.

In the preserved main text of FIAM (there is also a very detailed plan[1]) the problem is not solved and, like in AMR, it is posed again and again in ontological terms. At the beginning of ch. 1, Russell tries to distinguish pure numbers from terms, by considering the first ones as indefinable, in order to be able to present them as simple notions (or 'meanings'). Therefore, he now intends to put the whole/part relations behind numbers and even behind addition itself (even on pain of transforming it into a process of *synthesis* like that described in AMR). We are told about three stages: '(a) Pure Number (b) applied numbers and addition, involving terms, but not whole and part (c) addition and division, ratios and fractions, involving whole and part'. He justifies the priority of pure numbers with the argument that they are absolute as logical meanings and, therefore, of a similar nature to that of predicates: '"A is one" is formally analogous to "A is red"'. However, he soon realizes that the problems of such a strong distinction between terms and meanings lead directly to the need for regarding number as predicate and also as simple number, which is equivalent to admit (like Moore) two kinds of diversity: conceptual (based upon

---

[1] It is a long plan of the work of which only a few chapters, at most, were actually carried out, but it is interesting because it shows a global idea. Besides, it is detailed enough to contain useful arguments and distinctions. I shall often refer to it (under the term 'plan'), regarding it as a stage previous to the content itself of FIAM which has been preserved.

meaning) and merely numerical, although he seems later to preserve only the second one (more in agreement with Moore's pluralism). This situation shows, once more, that the problem is mainly an ontological one. In the subsequent analysis of addition (p. 231) he recognizes that a sum of mere units ('*A* and *B* are two') must be *prior* to the corresponding arithmetical proposition (1 + 1 = 2). It demands, consequently, an attempt to describe the true meaning of arithmetical addition while separated from numbers in themselves. All that brings us again to the ontological distinction between terms and meanings (predicates) and to an 'ultimate' type of addition based on a combination that would be 'indefinable'.

The problem of the priority between logic and mathematics is also not clearly solved in POM1.[1] Here the order of the concepts is very similar to that of FIAM, though the concept of *collection* is introduced at the beginning, together with pure number. However, Russell continues without deciding to regard logic as previous, despite that some of the undertaken analyses directly lead him to this result. Whitehead's outline again has the pre-eminence and seems to be difficult to fuse with Moore's ontology. I think this is perhaps the attempt where one sees in a clearer way Russell's failure to place Whitehead *1898a* into the framework of the indefinable terms.

The first two chapters are used to state the simple and indefinable character of pure numbers, although he insists on the need for resorting to the collections (or aggregates) of terms as an indefinable ground to the process of arithmetical addition. The argument is that numbers are not applicable to collections since these are mere additions and cannot define any number: numbers are pure concepts and cannot be constructed by counting, but through an intrinsic relation (*ratio*, like Couturat *1896a*), which transforms them into an ordered series. For this reason Russell thinks that this relation is the basis from which arithmetical addition must be defined, and he discards 'pure' addition and states for the first time the possibility of a simple, indefinable relation (among numbers) which is logically prior to addition.

But this view leads to the ontology of terms and their addition. That is why he later distinguishes (I/3) between arithmetical addition and a more 'philosophical' one referring to terms which is combined with the (ultimate) concept of ratio. The formal model will be: $(n) + (1) = (n + 1)$, whose meaning constitutes, according to Russell, a genuine *axiom*: 'The number applicable to a collection formed by adding one term to a given collection is the number next after (in the order obtained from ratio) the number applicable to the given collection'. Starting from this axiom he introduces the commutative law and a set of four axioms to rule the addition of integers.

It will not mean, nevertheless, the priority of logic; Russell avoids identifying this 'pure' addition of terms with logical addition, which appears much later (II/3), with regard to the whole/part relation (the source of the logical calculus), and is defined by the usual laws of the algebra of logic. Addition and multiplication will be, in this way, the ground on which algebra and arithmetic will be able to be regarded as 'coordinated' through the 'duality' that joins them, and serves as a foundation for the mutual replaceability of their basic concepts. That is to say, they

---

[1] In *1899a* there is a new piece of information: it is concluded that relations are the alphabet of logic and that they can therefore (together with other notions) give an account of mathematics. However this line of reasoning could only be confirmed when Russell could have an adequate logic available.

seem to belong to the same level of logical priority or, which means the same, any of them implies or presupposes the other: it seems to be the theory of the 'fusion' by Boole and Peirce (see 1.1 above), in the framework of Whitehead's ideas. Russell realizes the problem, as well as the *possibility* of making the logical calculus *previous*, but he takes the opposite direction (we have seen it) by basing arithmetical addition on the addition of terms, and even by accusing Grassmann's general addition of being a 'disguised' form of arithmetical addition.

The result is the definitive establishment of the two forms of addition: logic and arithmetic (which rests on addition of terms). As he is depending on Whitehead, he is forced to maintain logic as a branch of mathematics, although it is parallel in generality to the other branch (the one starting from number); as he is also depending on Moore, he is forced to try to found the two forms, respectively, on a sort of ultimate and indefinable 'ontological' addition (that of terms), and on a relation, also ultimate and indefinable (that of whole/part) which depends on the same ontology. Therefore the possible answers to the posed problems can be found only in the relations between logic and ontology: undefinability does not serve as a criterion for choosing between logic and mathematics as the true logically previous science. As we shall see, ontology will also be incapable of being entirely clarified through the search of the simples. In fact, a new and enormous difficulty will take place: the problem of harmonizing the belief in a true analysis with the criterion of the ordinary language (inherited from Moore) as guarantee to find the true meaning. Let us now go over the evolution of the relation between logic and ontology.

Russell's underlying ontology at this stage (beginning already in AMR) is an almost literal copy of that of Moore. However, there is a link with his preceding work (FG): he continues to be convinced that the method to be applied is the analysis starting from actual judgments of mathematics. The special feature proceeding from Moore is that, now, we have not only to describe and isolate the basic ideas presupposed in these judgments, but also to verify that the relations among the basic terms take place as well among the elements of *judgments in general*. This is because all logical judgments presuppose the ideas of identity, diversity and unity (and any relation): 'I shall, therefore, in the present Chapter, set forth briefly the irreducible types of elements, out of which I conceive judgments to be compounded. The various kinds of identity, diversity and unity will, I hope, emerge from this analysis' (AMR, I/1). This will be, I think, Moore's most important influence at this point: to have convinced Russell that undefinability must start from judgments in general (like in Kant and, to some extent, in Bradley); that is to say, from what we could call ordinary language.

Starting from this point the standard elements of Moore's ontology, all based on the relational theory of judgment, appear step by step. Thus we find (everything in AMR, I/1): (i) a criticism of the subject-predicate pattern because it cannot embrace all judgments; (ii) the definition of 'term' (equivalent to that of Moore's 'concept'): 'whatever can be a logical subject I call a *term*'; (iii) the acceptance of any idea as a logical subject: 'Every possible idea, everything that can be thought of, or represented by a word, may be a logical subject';[1] (iv) the criterion of classifying such logical subjects through their *difference of being*, as distinct from the difference as regards existence:

---

[1] Against Bradley, from whom the subject should be Reality as a whole (see my *1990g*).

'Existence is a predicate, unanalyzable as to meaning, which may or may not belong to a term. When a term has this predicate, it is called an *existent*'; (v) the application of this criterion to found a category of judgments: the existential ones, which are always secondary and incapable of serving as a ground to any theory of judgment. All that finally leads to a classification of judgments.

On applying the former theory to the subject-predicate distinction (I/2), a classification of predicates and terms is obtained, which underlies the entire theory of manifold, addition, and logical calculus. We must emphasize Russell's efforts to make Moore's relational theory compatible with a certain respect for predicates, in so far as these (and only these) can define classes by means of the notion of extension of a concept. Thus, Russell divides predicates into two kinds: as terms and as meanings, which respectively yield addition and synthesis; and terms into two further kinds: existents and non-existents (which will be called 'contents', still following Bradley). The important thing here is how logic and arithmetic are set up on a relatively independent ontological ground. As we saw above, the problem appears when the addition is constructed: on the one hand, an intensional theory of classes is proposed, since only predicates (as meanings) can define them; on the other hand, predicates are regarded as mere terms as well, which seemed to be the only way to maintain a relational pluralism.

That is why Russell is forced (in I/3) to introduce the notion of an 'assemblage of terms' as a mere collection of subjects; through this notion he shows how numbers cannot exhibit any other common predicate than that of being members of assemblages. On doing this, he becomes unable to think of constructing numbers out of classes in themselves; besides, he complicates the previous theory with a kind of entity difficult to classify as regards the duality between terms and meanings. On the other hand, he emphasizes the intuitive character of numbers by comparing them to colours, with the aim of strengthening their undefinability. In fact, Russell did not solve the problem of the duality of numbers, which seem to be predicates (concepts) as well as terms (units), until he was able to think of *logical* indefinables (once assimilated Peano's distinction between membership and inclusion; see 3.1 below).

In the FIAM plan almost all these ontological complications disappear, including the analysis of the various forms of judgment and the classifications of predicates and terms. That seems to be based on the drastic decision to place 'pure' number as a concept previous to the rest of mathematical concepts, starting directly from it for the analysis of addition. This has the advantage of directly leading to the whole/part relation as a ground, but, as the ontological status of the terms is undetermined, certain problems arise with the first analyses, in particular concerning the distinction between terms and meanings. This leads, as we saw above, to the undefinability of addition and to the doubt about what is truly previous, despite that the term-meaning distinction goes more deeply into ontological analysis, especially by the possibility of mutual reference (as can be seen when the distinction is applied to 'being', i.e. to *what it is*, which is both term and meaning).

These doubts lead Russell to consider, as hypothesis, some strange possibilities, like that of dispensing with concepts in order to eliminate the problems about the being-meaning distinction (FIAM, 11): 'A class is defined by a common relation to a common term; there is .. [therefore] no class including *all* terms. Terms are defined by *one* and *being*; nothing is added by *concept*.[1] It is an interesting possibility, once interpreted as ontological elimination. Russell's example is that of points: they are class-concepts, but they only appear as terms in propositions. When they presuppose a meaning (e.g. 'a point can be *punctual*') we should remember that they can be regarded as mere terms defined by means of their relations with other terms (e.g. when a point is defined by its distance to others), but not through that meaning. Russell tries to apply this conclusion in a general way: 'Similarly number would be defined by a peculiar relation to things numbered, which holds for all numbers, or even by a relation to 1. In this way propositions in which class-concepts occur as meanings may be abolished' (FIAM, 11). With that he tries to follow the line of Moore's drastic simplification, but this particular device also has all the merit of conceding the priority to terms through the elimination of meanings in favour of a relation. As we shall see below, here seems to be the seed of the reduction of properties to relations and therefore of the later principle of abstraction.

Through similar recourses, Russell achieves the following summary of his ontology in 1899, under the form of a list (which I simplify) of graded series of logical priorities (FIAM, 26): (1) substantives only as terms; (2) sums of substantives (*A* and *B*, etc.); (3) propositions with only one term; (4) propositions with several terms, i.e. assertions of diversity and numbers; (5) terms correlating to concepts not used as terms in (3) and (4) (i.e. the terms 'Being, Diversity and Numbers' in themselves); (6) propositions with several terms expressing relations among them; (7) relations as terms. Again we can verify the doubts we saw above with regard to the notion of priority. It seems that Russell is sure that there has to be a ground ontologically simpler than that of arithmetic, but he does not find the way to articulate this ground within his idea of logical calculus as subsequent to mathematics in general. He still needed a clear theory of relations in order to take advantage of its useful ontological implications.[2]

In POM1 we find the same logical order as in FIAM. Pure number precedes all the rest of the concepts; however there is the novelty of collections as an attempt to play the role of the former manifolds in general. There is also some progress in the theory of relations, which is already firmly established, we are told, as a guarantee of the pluralism needed in any sound mathematical philosophy. As I shall refer in detail to relations below, I shall only point out here that their ontological role was essential as an instrument to dissolve qualities (properties) in favour of *terms*, in the case of symmetrical relations, or as irreducible basic elements in themselves (having the

---

[1] Here he still makes use of Bradley's distinction between *what* and *that* since, Russell adds, if we identify concept and meaning, then such a distinction *cannot be maintained*: 'For *what* it is, is itself, and cannot be put into a proposition'.

[2] In *m1899a* this is somewhat clarified into the framework of an analysis of diversity. Russell denies that there are two types of difference, the material one and that of content, and he defends only one: the material diversity of content (which is already numerical), with the argument that the distinction between concept and existent is not clear and can only be maintained as numerical diversity: 'To be an existent is to be a term with a peculiar relation to existence, and to be different existents is to be different terms, each with this relation'. For this reason, the numerical difference is a difference between different terms (*m1899a*, C/4).

same importance as proper terms), in the case of asymmetrical relations (IV/3). This is perhaps the fundamental seed of all Russell's subsequent philosophy, in harmonizing the criticism of the subject-predicate ontology, the elimination of certain entities, and the kind of definition that would characterize the method in later works. A similar vision appears through the coincidence between logical and ontological analyses, which leads Russell to reject the theory of Hegelian tradition according to which the part is as complex as the whole (IV/4):

> Such a view can only be maintained by distinguishing logical analysis from division of substance, or of existent constituents... As I admit no distinction between these two methods of proceeding from whole to part, or rather, as I regard the latter as a particular and misunderstood case of the former, I maintain that every whole is logically subsequent to any of its parts, though not to all its parts together.

In what seems to be a discarded chapter (entitled 'Plurality'), which is included in the materials of POM1,[1] the kind of analysis involved in this vision of the whole/part relation is somewhat clarified. Here there is an attempt to introduce the concept of definability by means of that of ontological plurality. Russell admits two kinds of complexity: the first one is based on simple 'wholes' that can be easily analyzed into parts; the second corresponds to those wholes whose internal unity makes this division impossible. In this way he naturally arrives at Moore's distinction between simple and complex concepts. These latter always imply relations between constituent terms, but these relations are always subsequent to the fact itself of the plurality of components. Therefore the logical (ontological) order of concepts must necessarily start from the terms themselves rather than from the classes to which these terms belong: 'The fact is, plurality is prior to classes'. Starting from that, the most elementary class will be that of terms (which coincides with that of substantives), which has to be defined by extension. Similar arguments lead Russell to the view, which characterizes the kernel of POM1, that 'arithmetical propositions presuppose terms, and are not to be reduced to ultimate relations of numbers' (*ibidem*). However, as we saw, Russell did not decide to clearly establish a *pure* (or philosophical) addition as a ground for arithmetical addition, mainly because of the danger of confusing this pure addition with logical addition, which he needed to introduce, at this stage, as a different kind (under the influence of Whitehead). This was probably why he finally discarded the chapter.

The whole situation seems to be an internal fight between the influences of Moore and Whitehead. When the first one prevails, the indefinables tend to be unified in the framework of an 'ontological' logic; when the second prevails, two kinds of indefinables have to be admitted: those belonging to logic, which cannot be then previous to the rest, and those belonging to mathematics in general (to 'universal algebra'), which are to be identified with 'pure numbers' and, at the end, simply with numbers. That is why Russell cannot admit the definability of numbers, nor can he sufficiently clarify his notion of definability. The problem was that although, like for Moore,

---

[1] That could belong to FIAM, in whose plan there is a similar title: 'Divisibility and Plurality' (II/2), with a similar content (just as it appears summed up).

definability is always analysis into simple terms, for Russell 'logical' simples (terms) are not separated from 'mathematical' simples (units). For this reason he is forced to admit *indefinable complexes*, as in the case of any number other than one: 'Since 2 is one number, 2 cannot be definable; but 2 may nevertheless be complex. One complex unit is one which presupposes many units, and is thus not logically prior to all plurals' (*ibidem*). This leads him to admit that the *whole* differs from its components in an indefinable way: it presupposes them, 'and is determined by them, but not defined by them'. Starting from this view we can imagine Russell's surprise when he knew that numbers can be *logically* defined in a completely explicit way.

2.3. The evolution of the main concepts

As for *numbers*, for Russell they are indefinable ideas at this stage, which are capable of necessary connections constituting the basis of arithmetic, only from whose analysis we can apprehend them. This position would be maintained at least until the Paris Congress of 1900, in spite of the fact that sometimes Russell seems to suggest the possibility of some kind of definition.

In AMR he states that they are 'fundamental and simple [ideas], incapable of analysis or derivation, and not involving any prerequisites whatever' (II/I). As such ideas, they are the ground to arithmetic through judgments that connect them (e.g. addition) and offer the basis for an 'existential' application, which is always logically subsequent to those necessary (but synthetic) judgments. Consequently, 'if we are to say anything about number itself, we must analyze the necessary judgments of Arithmetic' (*ibidem*), and the study of these relations show that numbers are not properties abstracted from collections of terms, but logically independent contents (II/2), i.e. 'pure' meanings which we can know only through intuition (it would be impossible to explain them to someone who previously would not understand them). Therefore addition cannot give an account of them; it rather presupposes them: 'Each pure number is a unique indivisible content, not definable otherwise than as the number it is'. For this reason Russell states that in an arithmetical proposition of addition, the sign of equality expresses a necessary connection between logical contents and not an identity; otherwise the number expressing the sum would not make sense by itself and we should not have an actual addition (*ibidem*).

In FIAM it is worth only referring to the attempt of distinguishing between pure numbers and terms. We cannot, we are told, constitute the integers through the idea of 1 and the addition, because the latter already implies plurality, and because integers are not *sums*, even though they suppose some kind of addition: 'every integer is indefinable... This being the case, every integer is one simple idea' (FIAM, 1). Then Russell goes on to applied numbers, addition of terms, whole and part, and ratios, but achieving no solution to the problems referring to the distinction between being, meaning, term and concept, which was necessary to obtain that universal and philosophical mathematics that Russell pursued as the firm ground for logic and arithmetics.

In POM1 Russell insists again that pure numbers are independent, though terms (already in collections) set the ground for defining an addition of terms prior to other kind of addition. Numbers are *previous* in the sense of being presupposed by those collections (I/1), therefore the impossibility of defining them through addition is confirmed, and their undefinability is concluded. The possible objection that some numbers can imply other numbers (e.g. '2 implies 1') is refuted with the argument that if 1 appears as having logical priority it is due only to the fact that collections of a term are prior to the rest. The programmatic conclusion to the entire work is: 'there is therefore no reason to deny that all numbers are indefinable and simple. To prove that this is the case is obviously impossible: all that can be done is to refute suggested definitions. I shall therefore assume, in future, the simplicity of all numbers' (I/2).

As for *whole and part*, which is identified here as the logical calculus, the most important things have already been said (see 2.2 above); I shall add only some traits of the evolution of this concept in the manuscripts. In AMR the most important thing is the admitted correlation between whole and part and implication (I/2). This was not, of course, any 'technical' novelty, but it made it possible to regard the idea of 'to be a part of' (which supposes the non-distinction between inclusion and membership) as the basis of the logical calculus, and to permit that the parallelism existing between this correlation and that of extension-intension can be extended, until embracing as well the relation between addition and synthesis. In this way Russell managed to present classes through their defining predicates, and implication and the property of being a term (termhood) as defining predicates, respectively, of the class of predicates and of that of terms, which is another contribution to this rare mixture of grammar, ontology, logic and arithmetic which AMR is.

FIAM adds almost nothing to this view, except for the attempt of distinguishing between term and meaning, which, as we saw, leads to the admission of two classes of diversity, and therefore to make the sharp limits of applicability of the whole/part distinction disappear. However in the general plan of the work one sees how Russell thought to devote a very long part (with seven chapters) to whole and part as a ground for addition of numbers, plurality, classes and logical calculus, in what sometimes seems to be an increasingly pre-eminence of arithmetic over 'logic'. This trend seems to be stressed in POM1, where the concept of ratio comes to precede that of whole/part, once its capacity to construct the series of numbers (through the definition of order) is admitted. This leads Russell to see addition of terms as more important than arithmetical addition, and therefore to a clear loss of importance of the whole/part relation, which, although continuing as the basis of an entire part of the work (part II), is relegated to serve as foundation of the 'logical calculus'.

In this way the whole/part relation, although it is stated as 'a special ultimate and indefinable relation' (II/1), it lacks enough power to define certain fundamental concepts of mathematics, as 'any' or 'all'. Thas is because the whole/part relation would require a reference to *simple* parts in a finite number, while these notions are characterized precisely by involving the intensional reference to infinity (II/2). On the other hand, the foundation of logical calculus through whole and part can progress (II/3) in so far as it is made independent from arithmetic. With this aim Russell defines it as a transitive and asymmetrical relation capable of constituting a new whole coming from two given wholes, either as the sum of both or as their common part. If we add all the terms (the universe) and the equations defining the complement (and the null class), we have all the requirements for the logical calculus.

I come now to *order*. AMR lacks the chapter on ordinals (II/6 of the general plan), but resorting to the preserved materials it is easy to see that Russell's idea was to introduce it as a consequence of the application of the concept of *ratio* to numbers. Ratio (II/3) is introduced as the converse of multiplication and makes it possible to regard numbers as *relations* (between a collection and a relative unit) that can be neither reduced to, nor inferred from, intrinsic properties of the terms between which they take place. In the two next manuscripts order increased its importance until covering an entire part (each one with various chapters). Such a change was almost surely due to the influence of Couturat and Whitehead, without excluding the possibility that Russell became progressively convinced that without Cantor (the greatest supporter of the notion of order) it would not be possible to give an account of mathematics from a logico-philosophical point of view. Concerning FIAM only the plan of the work is available on this subject, though it suffices to see how order is presented for the first time as a simple and indefinable idea, which makes way for the various kinds of series (logical, numerical, of magnitudes, and of space and time) and also (so it seems) for the distinction of sign.

POM1 introduces a subject which enters in full in the corresponding part (IV) and that will characterize order in later works: asymmetrical relations. Here the various chapters suffer certain changes as regards FIAM; the distinction of sign is now presented immediately after the first chapter (the 'meaning' of order) and before asymmetrical relations, for which it is the foundation. Then the various kinds of series appear (where the whole/part relation replaces the order of magnitudes if we compare to FIAM), together with ordinal numbers (a general plan already similar to the definitive presentation of POM). The most important thing is to observe a change in the viewpoint toward a new vision of order as an independent and primitive idea. Russell even accepts the impossibility of defining continuity with no reference to order, quoting Couturat *1900a* as a support. Likewise, he recognizes the importance of the concept of order for irrationals, projective geometry and causality (which depends on the order of time). That is why he claims for the higher philosophical status for order. As the general position is already similar to that of POM, it suffices here to point out Russell's insistence on the indefinable character of order: 'the very meaning of the word excludes its reduction to any lower terms'. He adds also that, though order is based on a simpler and more general relation (difference of sense), this relation is nothing (as difference of sign) but the characteristic and ultimate trait of asymmetrical relations. In this way relations come to embrace order.

Concerning *quantity* and *magnitude*, Russell's position in FG already denoted certain Hegelian influences, which affected mainly the pluralistic character that common sense assigns to these notions and that Hegelianism denies (see my *1990a*). AMR lacks the crucial chapter on this subject,[1] but the Hegelian influence seems to be present to some extent, possibly because the work plan was already made precisely when Moore's influence began to take effect. On the one hand,

---

[1] Which should be at the beginning of the book II and be entitled, in a Hegelian way, 'Transition from Number to Quantity', a very similar title to that of his *1897c*.

we find 'transitions' from some concepts to others; on the other, some concepts are presented as the result of the union of the other two (in the style of the Hegelian syntheses). If we regard the preserved material it seems to be inferred that Russell intended to present quantity as the result of the application of the concept of ratio. From it, one would come to the idea of equality and inequality (both independent from number, which, along with order, would lead to the realm of pure quantity (I/3) (it seems with the argument that the relation of inequality leads to that of degree of quantity; I/4).

FIAM already sets the ground for what will be the part devoted to this subject in POM, and that, together with the material actually written, acquires a special interest given that the corresponding part in POM1 has not been preserved (possibly for having been reused in POM[1]). Let us look at it briefly. In the plan of the work magnitude is already established (against FG) as the concept with which a number of terms has the same *sui generis* relation (always referred to a given quality). In what was actually written we can find three practically complete chapters. From our methodological point of view, the most important thing is to observe how the relations between the introduced concepts are stated. Russell explains it as follows (in what seems to be the beginning of ch. 2): *magnitude* is any term greater or less than another, and *quantity* is any term with magnitude. Every magnitude is related in a special way to some term (of which it is magnitude), and two magnitudes are always unequal, whereas two quantities of the same magnitude can be equal. This position leads, after the examination of various possible theories (Russell's usual method for 'proving' the results from several hypotheses), to the reduction of the whole subject to two *indefinables* (every particular magnitude and the greater/less relation) and three *axioms* (that of inequality and asymmetrical and transitive laws).

This is doubtlessly a more 'logicist' theory than the reduction of magnitude to a common property of equal quantities (making 'equality' indefinable). Russell admits that this possibility would decrease the number of indefinables, whereas according to the adopted solution every magnitude is indefinable (FIAM, 8). In any case the offered solution has the advantage of replacing the existence of an abstract quality (or relation), which is necessary for postulating a subsequent 'equality', by a *magnitude* of this quality and its particular instances (the qualities) ruled by the axioms. With that a certain logicism remains guaranteed, at least in so far as nothing depends on number or measurement (FIAM, 10). As for POM1 there is, as I said above, no trace of the entire part III on quantity.

The last relevant main subject is that of *space* and *time*. Neither AMR nor FIAM came to a stage advanced enough to give an account of geometry or 'abstract' physics. The previous problems were so serious that these subjects had to wait to be dealt with in POM1 and, in a more definitive way, in POM. Something can however be conjectured by taking advantage of Russell's programmatic remarks. In the previous justification to the content of AMR it is said that geometry will come after dealing with extensive continuum and quantity as an application of number, and that it will deal with order, dimensions and distinctions of quality. The final stage would be

---

[1] Another possibility is that these almost four chapters, that seem to belong to the material of FIAM, really belong to POM1, for which there is important internal evidence.

reached, through other 'transitions' and by means of a new consideration of extensive quantity (which is identified with substance), with the notion of *thing* and the mathematical principles of generalized dynamics. We find a similar ordering of themes in the FIAM plan.

POM1 adds a very important reductive progress (in part VI, devoted to the subject), not by admitting points and instants as ultimate indefinable constituents, since this was the position already assumed in FIAM, but through the explicit claim that the analysis of the element common to time and space shows it as containing nothing which does not belong to the preceding concepts of number, whole and part, quantity and order. The reason is that the peculiar element of space and time goes beyond the mathematical treatment and it is, therefore, irrelevant (VI/1):

> It will thus appear that what mathematics considers about space and time (with the important exception of dimensions) has already been analyzed in the preceding sections, and is in no way peculiar; while what *is* peculiar is mathematically irrelevant (with the same exception), and is not à priori in any other sense than in which individual colours or tones are à priori.

This is an important position since it states in a clear way the general logicist tendency of POM (which, incidentally, can be regarded as another inheritance of the Hegelian tradition, already present from FG, according to which from pure concepts we must proceed, through several transitions, to 'matter' and nature).

Points and instants will then be the materials with which space and time will be constructed. As simple and indefinable constituents, they have no descriptive or quantitative significance. The concept of distance, as an asymmetrical relation in time (with its various spatial meanings), gives way to the definition of dimension (VI/4):

> a collection of terms has n dimensions, when many series can be formed out of the terms, each series being defined by some common quality; when, further, these qualities themselves can be arranged in series, each defined by a new common quality; and so on, until, at the nth stage, all the qualities involved form a single series.

It leads Russell to a somewhat conventional vision of dimensions in which *we* are who assign them by deciding which series will be granted as self-sufficient. Only in this way the number of dimensions comes to be a genuine property of multiplicities. Russell ends by offering a set of indefinables for geometry, in agreement with the ideas already introduced in his reply to Poincaré (*1899b*): distance, projective straight line, angle, projective plane, etc. (for details, see my *1990a*), which were subsequently analyzed and reduced to simpler logical components in POM. Finally a simplification is announced, though it seems that it was never written.

2.4. Concepts, axioms, presupposition and implication

Right at the beginning of the AMR introduction appears what will be, more or less, Russell's general position in this whole stage (although from FIAM he emphasizes more the concepts than the axioms): 'It is the purpose of the present work to discover those conceptions, and those judgments, which are necessarily presupposed in pure mathematics'. The first ones are indefinable, so that once they have been discovered, can only be indicated and their meaning intuitively apprehended. The second ones are necessary connections between different conceptions, and their truth, although also intuitively captured, does not depend on the fact that this apprehension does or does not take place (with which Russell emphasizes Bradley's realism as well as that of Moore). Their nexus to logic is the following: '[such judgments] form the rules of inference or, in a certain formal sense, the major premises of arguments which use the fundamental conceptions'. They can belong to two types: '(1) If a thing is A, it is B, and (2) If one thing has an adjective A referring to another thing, then the other thing has an adjective B referring to the first thing', and they are fundamentals because 'they are involved in the procedure in question', i.e. they are *presupposed* in mathematical judgments. Nevertheless, nothing can be adduced to justify their truth. It is precisely this feature of axioms that serves to distinguish the fundamental from other conceptions that, although they are also indefinable (e.g. red, sweet) they cannot be *fundamental*, but they seem to proceed from abstraction.

At the beginning of the first chapter Russell explains with some more precision the status of judgments (or axioms) giving them certain logicist connotations:

> It is intended, in the present work, to discover the peculiarities, and to lay bare the fundamental ideas, of the various classes of mathematical judgments. Judgments of number, of quantity, of order, of extensive continuity, of motion, and of causality, will be successively examined. In all such judgments, we shall find some special type of connection or relation, affirmed to hold between terms which may form elements in non-mathematical judgments.

A certain Hegelian flavour can also be discovered (like in many other places in the work) when it is affirmed that a basic triad is present in all mathematics: 'Like all ideas of relation and connection, mathematical ideas presuppose the three fundamental ideas of identity, diversity, and unity. But each of these ideas has different forms'. The object of the work is precisely to distinguish them and to characterize those that are relevant, starting from the analysis of the judgments in which they take place. In the material actually written of AMR this 'analytic' method does not bear too many fruits (like in FG); in fact only in the chapter on logical calculus (I/5) a reference is made to the fact of *isolating* the postulates of this branch of mathematics. However this attempt is not achieved because Russell does not clearly state the *priority* of logic over arithmetic (or vice versa). What proves to be clarified is the impossibility of characterizing the fundamental conceptions without making reference to the existent connections between them, i.e. to the axioms, and vice versa: the impossibility of constructing such axioms without containing those intuitively apprehended conceptions. It is just this possibility of an immediate access to them that makes their relations have an intuitive basis as well. (That is why *presupposition*, namely logical priority, is the ground of the whole system: the axioms, just like the conceptions, can imply to one another; see 2.4 below.)

The title itself of FIAM (*Fundamental ideas and axioms of mathematics*) already gives the aim of the work. Unfortunately there is no introduction that sets out (as in AMR) the features of the attempt. Nor does the general plan inform us of more than the order and the peculiarities of the various subjects, but not of the particular traits of the ideas and axioms involved. The most outstanding thing is perhaps the effort in presenting numbers as indefinable essences, and the doubts about the ultimate priority between logic and mathematics. We are told (I/1) that every integer has to be considered as indefinable and therefore as a simple idea, which leads to recognizing that none of them is logically prior to the rest. However this leads (as we saw in 2.2) to distinguishing between term and meaning and, since concepts can play the role of terms as well as of meanings, the status of the intended 'fundamental ideas' (which necessarily have to be concepts) remains quite burdened with the entire ontological problem. For, though intuition will permit us to *recognize* simple ideas, these ideas only take place in ordinary (mathematical) language, under the form of connections whose truth we also intuitively recognize. All that requires however a process of distillation in search of those ideas (or axioms) that are presupposed *in all the rest* (though those not being simple also seem to be so from an intuitive viewpoint). That is why Russell does not want to leave any presupposition without the relevant treatment, which leads him to consider every idea in relation to all the others in a tiring and scarcely fruitful fight against language.

Starting from that, Russell's practice does not coincide with his claims that concepts are simple in themselves or they are nothing (in a way independent from our wishes). The task of philosophy is precisely to discover these indefinable terms (FIAM, f. 8 —of the second series of folios). Besides, since the logical status of the axioms is unclear, it does not seem possible to distinguish them from the mere rules of inference; and, since such rules are reduced to *implication* (FIAM, 27 ff), which is used as an intuitive criterion of logical priority, it is impossible to regard the method not even as minimally algorithmic. The only clear example of a set of indefinables and indemonstrables in FIAM appears as a conclusion to the pages devoted to quantity and magnitude, but as we saw the first ones are doubtful and the second point out properties belonging to many other relations (e.g. asymmetrical or transitive).

POM1 lacks also an introduction and a list of objectives, although these seem to be similar to those from the former manuscripts. Concerning axioms, we find only one set of them actually mentioned: that ruling the arithmetical addition of integers. I summarize it to show a piece of the kinds of Russell's results at this stage (I/2): (1) any two collections can be combined to form a third one; (2) if the first two have a number, so has the third; (3) numbers have an intrinsic order obtained from mutual relations (ratios); (4) the collection formed by adding a term to a collection of *n* terms has the number next after *n*. Finally, it is added that all the particular properties of addition can be deduced from the axioms. It is to be observed that, although it is true that the former list of axioms seems to summarize the entire process of analysis that Russell carries out in the previous chapters, however, the axioms have rather a factual character (apart from the second one, which seems to be a 'rule of formation'), very far from the rules of inference claimed in AMR as the paradigms to follow. Besides, they not even incorporate the usual properties of relations (like those offered in FIAM), which seems to coincide with the progressive trend to mathematization that we observed in preceding sections.

The problem is complicated if we regard the relation of the axioms to the principle of conceptual economy (the future Ockham's razor). The axioms seem to be offered *at the end* of the process of investigation, i.e. as a result, but this seems to contradict their 'indemonstrable' character, and proves that they have nothing to do with anything similar to an 'axiomatization' (which would *start* from a set of axioms in order to present the further propositions as theorems). Instead, we find an odd mixture between concepts and axioms, where all of them are gradually distilled through the process of analysis, and where Russell 'tries' with several indefinables until finding the most satisfactory combination for the momentary objective.

Concerning *indefinable ideas*, we have already seen above the order and the features of the main ones. It is to emphasize as well the same trend to mathematization referred above, which can be observed by simply remembering the priority of number over other ideas (though, as we saw, with some doubts concerning the role of collections and of logic in general). Let us consider, to finish the section, the evolution of the notions of *presupposition* and *implication*.

There are only two attempts to explain these concepts in AMR and they are very incomplete. According to the first one (I/3), the logical contents only acquire their predicates through relations, so that when we want to define (or classify) one of them, we have to regard the defining content as the result of a synthesis. Thus, we can say 'man is a rational animal', or 'red is a colour', but not '3 is a type of number', since given its character of an absolutely particular content, '3' is 'incapable of being implied by other similar contents'. Consequently, it seems that the relations of implication can only 'point out' simple logical contents when, and only when, the implying concept (the 'antecedent') is complex and therefore capable to be defined just by means of its implications (or presuppositions); which seems to mean that what is complex is presupposed by what is simple but not vice versa. This is the reason why Russell thought (as we saw above) that symbolic logic must be based on the mutual relation between addition of terms and synthesis of predicates (i.e. ontologically, between terms and meanings).

The second explanation starts from this relation and, through the evidence that intension and extension (the respective bases of synthesis and addition) are parallel, it shows the correlation existing between implication and the whole/part relation (AMR, I/4; the emphasis is mine):

> The axiom of intension is: If a, b, c be three predicates, such that a implies b and b implies c, then a implies c. The axiom of extension is: If A, B, C be three assemblages, such that A is part of B, and B is part of C, then A is a part of C. Let now A, B, C be three classes defined by a, b, c. Then the axiom connecting intension and extension is: If a

implies *b*, *A* is part of, or exactly coextensive with, *B*; or, conversely, If *A* is a class which is part of a class *B*, then *a* implies *b*. *Thus the two ideas of whole and part, and of implication, are exactly correlative.* Any judgment concerning the relations of classes allows us, therefore, to infer a judgment concerning their defining predicates, and vice versa.

With that Russell comes to establish a link between inclusion and implication, which depends (as we shall see later) on the non-distinction between implication of terms and implication of judgments, on the one hand, and on the non-distinction between inclusion and membership on the other. From this point of view the important thing is to realize that both analysis and definition have to start from this axiom, which is at the same time a logical, linguistic, and ontological one. By the way, Russell clarifies another side of implication by telling us that when two ideas (or conceptions) are correlative, *they are mutually implied* (in spite of being independent), which is equivalent to say that none of them is logically prior to (or is presupposed by) the other.

The second attempt (FIAM) contains a first explanation of the conditions of implication with regard to inference in general. The basic ideas are the following (FIAM, 27-33). First of all, Russell intends to establish a parallelism between arithmetical propositions and rules of inference: 'Arithmetical propositions —as $1 + 1 = 2$— seem in the same position, with regard to terms, as rules of inference concerning relations' (f. 27). This is important because it shows how Russell continues to depend essentially on his original goal (to analyze the judgments of mathematics) as much as on the influence of Whitehead (in the sense that the essential and common thing in both logic and mathematics is addition). On the other hand, rules of inference are presented as subsequent to their applications, under the argument that the propositions they necessarily connect 'seem all prior to their necessary connection'. Therefore, these rules are peculiar relations between propositions, usually between two of them and a third one: 'Thus "*A* is one and *B* is one together *imply A* and *B* are two". *Imply* expresses a unique form of relation' (f. 27). Consequently, the relation to the simple/complex distinction is the following: the premises and the conclusion should always be previous to their connection, if what is simpler is previous to what is more complex (f. 28).

As for truth, a certain reduction to truth values is attempted. In the case that a proposition (Q) has several premises ($P_1, P_2, \ldots P_n$), the following possibilities about these values are established: (1) if Q is T[rue]., $P_1, P_2, \ldots P_n$ are all T.; (2) if Q is F., $P_1, P_2, \ldots$ or P_n are F[alse].; (3) if P_n is F., Q is F.; (4) if $P_1, \ldots P_n$ are all T., Q is T. This *non-coincidence* to the modern material implication (Russell still did not know Frege or the relevant works by Peirce) is summed up in this way: 'If P_n is a casual one of the premises, Q's truth implies that of P_n; P_n's falsehood implies that of Q. The truth of Q and P stands or falls together'. Therefore, he concludes, somewhat unsatisfied, that the logical connection 'seems to be very complex'. But Russell was not very convinced of the reduction to truth values and he doubted about the possibility of regarding implication as an *ultimate* form from which the applications in terms of truth values were only inferred (FIAM, 28):

'If A is true, B is true' implies 'If B is false, A is false'. But 'If A is true, B is true' seems to mean 'A implies B'. Here A and B are propositions. 'A implies B' seems to be an ultimate form of proposition, from which is inferred 'A's truth implies B's truth' and 'B's falsehood implies A's falsehood'.

This is another sign of the essentialism of the 'true meaning' and of Russell's problems with the analysis in terms of truth values. In any case the progress that Russell's position supposed with regard to Moore's obscure notion of logical priority is obvious, and this progress is possible only through the notion of truth. From this point of view, Russell comes even to criticize Moore: 'there is a vicious circle in Moore's account of logical priority. "A implies B" cannot *mean* "A's truth implies B's truth"; for here a simpler case of implication is explained by one which is more complex'. The following objection, proceeding from a better approach to the truth analysis, is even more decisive: '"A implies B" *implies* "A's truth implies B's truth" and also *implies* "B's falsehood implies A's falsehood"'. Russell infers that the important thing about implication is to regard A and B simply as propositions, i.e. independently of their truth or falsehood. That is why he can draw the important conclusion that *implication can also take place between concepts*, and he can state the basic relation among definability, simplicity, implication and presupposition: 'Since "A implies B" has no essential reference to truth and falsehood, it is possible for simple concepts to imply others. "2 implies 1", "Being implies substantives". Since "A *is* implies A" where the first term is a proposition but not the second' (FIAM, 29).

However, the actual progress appears mainly through the relation between implication and logical priority: 'If we have any proposition "B implies A", we say A is logically prior to B' (FIAM, 32). From that Russell cannot avoid drawing the conclusion that logical order (i.e. the whole possibility of defining or reducing through simple constituents) depends on implication (apart from mutual implication, which 'breaks' logical order). This is the reason why Russell regards numerical order as a perfect logical order: '"A and B are two" implies "A is one" but not viceversa... We have here an ideal case of logical order'. To sum up the position of FIAM: if A implies B, then B is prior to A; and since logical priority is the converse of presupposition, to say that B is prior to A is to say that B is presupposed by A. Finally, it also seems to be inferred (especially by regarding implication between concepts) that what is logically previous exhibits a greater simplicity than what is subsequent.

We come now to POM1, whose explanation supposes the clarification of some obscure points. The following statements are maintained and justified (in II/1, unless otherwise indicated).

Implication is a fundamental and indefinable relation characterized by the transitive and symmetrical (or asymmetrical) properties, whose converse is logical priority. When implication is asymmetrical, 'we say that *a* is prior to *B* when *B* implies *A* but *A* does not imply *B*'. When implication is reciprocal, Russell says that the propositions involved are coordinated.[1]

Implication is the ground for all reasoning: 'in order to infer *B* from *A*, we must know that *A* implies *B*, as well as that *A* is true; indeed "*A* implies *B*" is a type of proposition containing the logical essence of what is called inference'. As for truth values, his position is the same as that of FIAM: if *A* implies *B*, then *A*'s truth implies *B*'s truth, and *B*'s falsehood implies *A*'s falsehood, though 'the implication tells us nothing as to whether *A* and *B* are true or false'.

Simple terms may be implied by propositions. This assertion has clear ontological commitments and points out the whole/part relation: 'when the proposition implied is of the simple form "*X* is", we may say that the implying proposition implies the term *X*; for if any term *X* can be implied, then that term is an entity. Thus simple terms, which are not propositions, may, by this extension, be implied by propositions'. Russell adds that the extension is basic if we want to identify logical priority to the relation from the part to the whole, since a whole can have simple parts.

Thus, *the general criterion of simplicity depends on the relation of implication*, for both propositions and terms. Russell does not explicitly state the possibility that a term implies another (as he did in FIAM), but he regards it as granted in saying this general criterion is useful with propositions just in so far as we use the constituent terms as a *sign of simplicity*: 'Wherever we have a one-sided implication, what is implied is simpler than what implies it' (we shall see below this is also the case with regard to definability). Russell offers three examples: (i) Euclid's axioms are simpler than his propositions, which imply them; (ii) if '"Socrates is a man" implies "Socrates is mortal"', 'it is evident that the latter proposition is simpler than the former, for *man* is a concept of which *mortal* forms part'; (iii) if we have a proposition that affirms some relation between two entities *A* and *B*, it is evident that the proposition presupposes these entities as well as the relation itself, and that they are all simpler than the proposition. Consequently, Russell can use logical priority (the converse of implication) to characterize the concept of 'whole', on the ground that any complex term presupposes the *being* of the simple terms composing it (its parts): 'Between a whole and any of its parts, there is a specific simple relation, that of logical priority or of whole and part'. All this leads us to the well known distinction (which we saw in 2.3) between aggregates and units, i.e. between wholes that can be specified through enumeration (analysis) of their parts and those that cannot (e.g. propositions). Of course the latter connection between implication, logical priority, simplicity and the whole/part relation completely vanished when Russell learned from Peano the distinction between membership and inclusion.

However the whole/part relation is not identical in meaning to logical priority, despite that they are closely related. This also contributes to clarifying the difference between implication and logical priority (which are not only converse): 'Logical priority is not a simple relation: implication is simple, but logical priority involves not only the proposition "*B* implies *A*", but also the

---

[1] Russell gives the example of the Euclidean axioms, which, except for that of parallels, are *prior* to proposition 32, since it implies them, whereas it is not implied by them; on the other hand, the axiom of parallels is coordinated with this proposition because the implication between them is reciprocal.

proposition "*A* does not imply *B*'". Though this would be achieved in the case that *A* is a part of *B*, it is also necessary to regard the whole/part relation as simple and different from any possible relation of a whole to another not being a part of it, which *would not happen* if the relation was that of implication: it would be necessary that the whole/part relation was always asymmetrical.

Russell's basic difficulty is going to be the possibility of reciprocal implication (already considered in FIAM), precisely *because this possibility threatens the reductive order* he wishes to achieve for basic concepts and propositions, as well as the assimilation of logical order to the whole/part relation. Russell inserts reciprocal implication into the type of relation whose terms remain affected in the same way by them, as when we say '"*A* implies *B*" implies "*B* is implied by *A*"'. In these cases it is said that the difference is only grammatical, but other cases cannot be solved in such a way (IV/4):

> Take, for example, the formulae for the angle-sum of a triangle and a polygon (Euclid I,32 and Cor.). The former is a particular case of the latter, and is therefore implied by the latter: yet in Geometry it is the latter which is deduced from the former, and thus the implication is reciprocal.

The problem is of a general kind, for it affects the whole deductive chain of reasoning: when two propositions are *equivalent* Russell thinks that the order of the corresponding chain of propositions related through implication is broken down (without forgetting that the elements of this chain can also be concepts). (And this order is *intrinsic* because it presupposes objective relations of simplicity-complexity.) To illustrate it Russell again resorts to the example of Euclid's axioms and propositions: axioms (unlike in FIAM) imply propositions and (for the majority of them) vice versa: 'The axiom of parallels is certainly distinct from the proposition concerning the angle-sum of a triangle: yet these two imply one another. Both must be true, or both false'. Through this position Russell wants to point out again that reciprocal implication *does not produce identity*, for the intuitive features of the two propositions involved are deeply different. The conclusion is the breaking of the logical order: 'Thus there is no logical order between them: either may equally be used to prove the other'.

Therefore, if Russell continues to identify logical order with whole/part order (the only one compatible to Moore's somewhat simplified logic and ontology), he has to take as paradigms only those cases in which implication is one-sided. To begin, he gives the example of the axioms common to Euclidean and non-Euclidean geometries, which, though implied by *all* Euclidean propositions, however do not imply by themselves (according to Russell) more than the first 28 propositions. The reason is that when a single proposition implies many others, then, if all the latter are true, the former is also true, whereas one of these isolated propositions cannot imply, as a rule, the former proposition. In the last analysis, the true (one-sided) implication can works only when there is a logical and ontological order from that which is complex to that which is simple. Thus, the relation of implication is also the one from that which is complex to that which is simple, so that if a proposition implies others, it is (as a rule) because the former is more complex than some of the latter (i.e. it has more parts). On the other hand, when that proposition is not more complex it is (as a rule) because implication is reciprocal.

As a result Russell concludes that logical order and whole/part order are the same (except for this case: $A > B, B > C \therefore A > C$, where the conclusion is not more complex than the premises and does not imply them). The consequences of such a correlation will be very important for his general method, in stating a parallelism between reduction (analysis) and presupposition. From an epistemological viewpoint this involves, according to Russell, a complete coincidence with the main object of philosophy: to discover the implicit (conceptual and propositional) presuppositions in all fields of knowledge. Since it is not true that the knowledge of a proposition implies that of its premises, then the most difficult task of philosophy is to discover the premises involved by the accepted propositions. This leads us to the last point of the section: the relation between implication and definition. Russell completely clarifies the connections as a consequence of what has been already stated: 'simple terms, having no presuppositions, are indefinable. And with complex units, the same seems to be true'. In fact, though a complex unity has presuppositions, it is different from them and the mere enumeration of its parts would not be equivalent to its definition (ch. 'Plurality').

2.5. The contradiction and the infinite

Already in the introduction to AMR Russell relates the contradiction 'of relativity' to the indefinables; that is why this contradiction embraces all mathematics. The problem is considered just after giving a general vision of the fundamental ideas of mathematics, by following Moore and his defence of pluralistic diversity:

> these ideas will be found for the most part indefinable, but their meaning will be so far as possible, indicated, and their fundamental nature proved, as they arise. It will be found that one pervading contradiction occurs almost, if not quite, universally. This is the contradiction of a difference between two terms, without a difference in the conceptions applicable to them. I shall call it the contradiction of relativity. This, with addition and the manifold, appear to define the realm of Mathematics.

Later on he goes more deeply into the significance of the contradiction, which certainly appears in all places where an element (simple or complex) can be regarded through different conceptions, especially when the whole/part relation is applied. This relation allows us, on the one hand, to see the whole as a part, and this, if it is not absolutely simple, as a whole (and in general classes as *one* or as *many*) and, on the other hand, to see the relativity of both notions, which can be understood only through a mutual reference (the part cannot be without the whole and vice versa). However, since the whole/part relation underlies addition, which rules all mathematics, relativity is extended to all its field. This relation (AMR, III/2):

> is one which seems to depend upon material identity and diversity, so that the sum of the parts *is* the whole, i.e. is the very same thing as the whole. Thus the relation of whole and part underlies addition, and hence all Mathematics. The relativity of this notion, by infecting the unit, leads to fractions, infinity, and the number-continuum.

In this way, though Russell states the contradiction still in somewhat Hegelian terms, he continues to maintain, with Moore, that analysis (the ground for definability) works necessarily through the whole/part relation, which makes the notion of *unity* possible, no matter how relative it is. Russell avoids to drawing the conclusion that simplicity is always relative only by means of a *contradictory* application of the whole/part relation; that is why the contradiction is unavoidable. The notion of simplicity is also affected: on the one hand, the complexity of a term depends on the conception applied to it (as we have seen); on the other, there are terms that cannot be regarded as complex: 'Any term, then, is more or less divisible, according as the conceptual difference, between it and the terms whose sum it is, is more or less' (*ibidem*).

In FIAM what is anticipated in the plan is only very partially developed. In particular, the infinitist ground of the antinomies is specifically claimed: 'Mathematical ideas are almost all infected with one great contradiction. This is the contradiction of infinity. All antinomies, I believe, so far as they are valid at all will be found reducible to the antinomy of infinite number' (IV/4). But the problem of continuity, zero and infinity, is reduced in a pre-Cantorian way, to the existence of the three respective axioms.

In POM1[1] the global situation becomes clarified, especially because the implications of both notions (the contradiction and the infinite) become explicit. The contradiction of relativity is seen now, through some progress in the theory of relations, as the result of the application of a defective logic, which cannot be overcome unless external relations are admitted. The antinomies of infinity and continuity are concentrated more and more in transfinite numbers, which, though they are not accepted, contribute to making an alternative Russellian theory possible.

The contradiction of relativity appears, we are told (POM1, IV/3), when the theory that all relations can be reduced to adjectives of the related terms is accepted (i.e. when we use the subject-predicate pattern); but there are cases (e.g. asymmetrical relations) where the adjectives corresponding to the relation cannot be obtained. Then the contradiction appears: it consists of 'a difference without a point of difference, i.e. without any difference as regards the predicates of the term which are different'. The argument starts from a certain type of asymmetrical relations (like inequality) where the relation produces different adjectives (in the case that we attempt the

---

[1] In a manuscript folio (*m1900a*) another similar contradiction appears: here identity is presented: as a relation (*A* is identical to *B*); as not being such a relation (two terms cannot be identical; if they were so they would be *one* and one term cannot be identical, since what would it be identical to?; as nothing (diversity would affect any two terms); and as something (the preceding position leads us to a vicious circle).

reduction to adjectives of the terms). However (as a rule) a relation, when analyzed into adjectives, assigns only those properties or qualities that terms already have by themselves, so that the relation is a *relation of adjectives*.

However, in asymmetrical relations the adjectives are different and cannot be regarded as the ground of the relation, since they are created by the relation itself; therefore we have here *a relation of things* and not a relation of adjectives of these things: 'in such a case we have, in both terms, adjectives of relation, but we have not a relation of adjectives'. After analyzing other examples Russell concludes that these relations give 'a conception of difference... without giving, apart from this conception of difference, any difference of conception'. The contradiction is present in all kinds of relations, so that it can be avoided by denying one of the implicit premises: that all relations have their ground in the related terms in themselves. By admitting 'external relations' the contradiction vanishes and the merely symmetrical relations can be 'reduced' to identity of content, with which Russell obviously progresses towards the principle of abstraction (see 2.6 below).

The part devoted to continuity and infinity in POM1 (V) supposes a rejection of Cantor under the general argument that his transfinites do not offer any solution to the philosophical problems involved (see 2.8 below), so that they cannot eliminate antinomies. However, Russell adds, they suppose a mathematically valuable analysis of continuity and bring up the essentials of the technical problems involved. Russell's alternative consists, briefly, in *characterizing* the infinite simply by saying that there is no number of all finite numbers; with that he thinks he is destroying Cantor's ground and allowing the possibility of talking about 'all finite numbers' without any fear of the contradictions in which we would fall into in admitting a particular number of all these numbers. However, in recognizing that Cantor's definition of continuity (see 1.2 above) better avoids the problem than his own alternative, and in admitting that Cantor's construction of transfinites creates a new branch of mathematics that enlarges traditional arithmetic, Russell paves the way for the later acceptance, which will happen after 1900 (see 2.8 and 4.1.3 below).

The progress as regards prior stages is, then, obvious. On the one hand the contradiction, which had important ontological roots, is solved through a technical device based on asymmetrical relations; on the other (as we shall see), there exists a trend toward a 'logicist' arithmetization, which will only culminate with the acceptance of Cantor. In both cases the ontological implications are not eliminated, but they are inserted into a new framework where extensionality and atomism will come to fruition. In this way Russell leads external relations much further than Moore ever thought and, at the same time, goes far away from Bradley.

2.6. Relations and the 'principle of abstraction'

As we have just verified, the importance of relations is increasing in the manuscripts. This is the reason why Russell makes an effort, not only in using them, but in building a classification of this field of the new logic, practically 'invented' by himself in *m1899a*. This classification allows not only the subsequent increase in the efficiency of applications, but sets the ground for one of the basic methodological recourses of POM: the principle of abstraction. With this approach and helped by Bradley's ghost[1] (which characterizes this entire stage of 'liberation'), Russell manages to dispense with 'properties' in favour of relations and, on this same line, to give an account of order. Likewise Russell will now fight (once again following Moore) to give the pre-eminence to 'numerical' diversity, which is more useful for the incipient logical atomism than the Hegelian diversity 'of content', i.e. to the pre-eminence of extension over intension (despite the gradual acceptance of Cantor).

In AMR this last feature is clearly visible, especially in his theory of manifolds (I/3). To create a class we need, according to Russell, only the unity of a logical subject, but not necessarily a defining predicate. In case the relation of a member to its class does not imply existence, as for instance occurs in the relation of '3' to 'number', this member ('3') is not a quality (it does not admit degrees) but, like all numbers, can be predicated of assemblages. This same possibility can be regarded as the possession of a predicate common to all numbers, but, since exhibiting this property means to apply a number to an assemblage of terms, then 'this property consists in a relation to terms which are not numbers'. With that we already have an example of the way to reduce a property to a relation, an important trait of Russell's later method. However, what is important now is to realize that the example can be enlarged even to 'existents', like colours for instance. A particular colour is related to the common predicate 'colour' in the sense that it consists only of relations (though in the case of colours they are true predicates), since: 'apart from relations no predicates common to all colours can be found'. Russell concludes that, though the 'logical contents' taken as terms can only belong to a class through the attribution of a predicate, on the other hand they can only acquire predicates through relations, with which, since external relations are involved, relations are regarded as terms.

In this way Russell manages to show (as I pointed out above) that although a class, unlike a mere assemblage, receives its unity from the possession by its members of a common predicate, however, on regarding the predicates themselves as terms (as in the example of numbers or colours), the common predicate should be a common relation to some other term, and this other term can be a predicate, a content other than a predicate (see 2.2 above), or an existent. Thus all predicates of a given thing form a class and have in common the predicate 'predicability of the given thing'. Therefore, although we can continue to defend 'identity of relation' as a sign of an

---

[1] I am referring to the 'supposed' Bradley who, according to Russell, had defended 'internal' relations and denied 'external' ones (I explained this point in my *1990g*).

intrinsic common predicate, we cannot discover it. In this way Russell not only plants the seed of what will later be one of the lines towards the principle of abstraction but also achieves the identification of the concept of manifold to that of the extension of a concept (which leads him to addition and logical calculus). Nevertheless, AMR lacks an explicit treatment of relations.

We do not find anything in FIAM but scattered remarks, although some of them are very important in illustrating the way in which Moore had defended external relations (in order to keep a true diversity between the related terms), and a genuine ontological status for the relations themselves. In the following I shall study some of these remarks.

The influence of Moore's relational analysis of judgment very soon appears: 'In order to get a proposition which may be true or false, according to the nature of the terms, we must have 2 terms and a relation' (FIAM, 19). Thus, a series of things can be stated for all simple terms: that *they are*, that *they differ*, etc. We could even identify two of them (A is B) still without getting a proposition, but only the attribution of the same concept to two signs. However, just when we consider *relations*, something else that does not belong to the nature of the terms is inferred: 'this is why relational propositions are synthetic' (*ibidem*). If a relation between two terms occurs, this relation will not take place between all the possible terms (unless it is diversity itself). Therefore, 'it is in relation that the particularity of concepts appears' (*ibidem*). If we have propositions of the type 'A is' or 'A_1 differs from A_2', it is irrelevant to consider what kind of concepts are involved. However, as soon as we consider the relations themselves 'they always involve a knowledge of the particular nature of our term' (*ibidem*). Hence relations are synthetic and, on the other hand, terms only show their nature in them.[1]

Later on we find the 'conclusion' to all this (FIAM, 26). According to Russell, diversity (the only true symmetrical relation) must be included among numbers since it only states a plurality. However when we proceed in this way, we obtain only asymmetrical relations as such relations, i.e. those presupposing an order making the reciprocal substitution of their terms impossible: 'Then we may say that relational propositions are distinguished by the necessity of *order* among the terms'. Therefore, no given relation will take place between any pair of terms: a specification of them will be required. Relations select terms, therefore they are external (*ibidem*):

> This seems to give the source of the talk about external relations. Some terms can have to others a given relation R, others cannot; these can form a class, and are supposed to have something in common —i.e. a common predicate. Their *nature* is supposed the source of the relation. But this *nature* can only be shown in relations. It is wholly contained in the concept itself, and does not lead to propositions without other concepts.

It is then impossible to regard relations as predicates of the related terms: we already do not need to think of anything that all members of the class exhibit.

---

[1] Russell goes further on making relations prior to terms; as we infer their existence from true propositions with the form ArB, in false propositions we can, likewise, carry out the inference: 'this *seems* to show that R is prior to ArB' (R being the *term* corresponding to the relation r).

There is still another reference to relations in FIAM. Besides, it is very interesting because it clearly states the elimination of symmetrical relations by means of the principle of abstraction (though still without mentioning it), so that it points out a connection with definability. The question is posed with regard to magnitude (in the final loose folios), in particular on whether magnitude has to be regarded as a common property of equal quantities (making equality indefinable) or if, on the contrary, it is better to regard each particular magnitude as indefinable (defining equality as 'identity of magnitude'). Russell finds the solution in his logic of external relations, by reducing symmetrical relations (the ground for common properties) to independent terms, and by making the asymmetrical ones irreducible (FIAM, 8 f):

> Many philosophers would decide this question off hand, by means of a general principle. Wherever two terms have the same relation to a third term, they would say, this shows that the two terms have a common property. But this doctrine is rejected by the logic adopted in the present work. A common property, according to the view here taken, is merely a third term to which the two given terms have a relation of a specified type.

Therefore the latter type of relation cannot be that of symmetrical ones. According to Russell, these are ruled by the symmetrical and transitive properties (from which the reflexive one follows), so that they are not genuine relations (no term can be related to itself) but rather 'sameness of relation to some other term not of the class in question'. By applying this conclusion to the problem of magnitude (as we saw above) Russell chooses the interpretation that makes particular magnitudes indefinable (which will be the solution adopted in POM).

Before entering into POM1, I shall consider a very important unpublished essay: 'The classification of relations' (*m1899a*). It proceeds from the same year as FIAM, but here Russell establishes the ground of his later theory of relations. I shall consider in the following the main three methodological points: (1) the classification itself; (2) the defence of just one type of diversity; (3) the irreducibility of asymmetrical relations.

(1) The classification states four classes of (dual) relations, according to whether or not they possess the symmetrical and transitive properties, which are employed as names: (i) *symmetrical*: 'if ArB, then BrA, and if ArB and ArC, then BrC' (e.g. equality, simultaneity, identity of content); (ii) *reciprocal*: 'if ArB, then BrA, but if ArB and BrC, it does no follow that ArC' (e.g. inequality, spatial separation, diversity of content); (iii) *transitive*: 'if ArB and BrC, then ArC, but if ArB, it is false that BrA' (e.g. whole/part, before/after, greater/less, cause/effect); (iv) *one-sided*: 'relations of which neither of these properties hold' (the only example is the subject-predicate relation, in the twofold way of predication and occupation of a time or place, although later on Reject would reject it as a true relation). From this classification, Russell 'formally' infers, against Bradley's 'ghost', that it is not true that all relations are to be reduced to adjectives of the related terms: if it were so, one would have to deny the latter two classes of relations (since both are asymmetrical). His parallel 'philosophical' inference is the next point.

(2) The most important thing in Russell's reduction of the two supposed kinds of diversity (the material one and that of content) to only one kind (the numerical one) was already seen above.

The argument was based, following Moore, on not admitting the distinction between concepts and existents (i.e. in rejecting the subject-predicate distinction); between existents there is the same diversity as between concepts, and this diversity is a numerical difference (which therefore implies plurality). Thus, by claiming that an existent is nothing but a term (and any logical subject can be a term), numerical difference takes place between different terms, i.e. between terms of only a kind. In this way, Russell states a link between: (i) Bradley and his criticism of the subject-predicate distinction (together with his acceptance of the theory of 'logical meanings'); (ii) Moore and his incipient logical atomism (based on the identification of things and concepts through the implicit distinction between *being* and *existence*); (iii) the importance of numerical diversity (this also was based on Moore, but Russell assumes it in order to use it in his theory of relations and, especially, in his pluralistic philosophy of mathematics).

(3) Russell's main conclusion is that no relation is equivalent to a pair of predicates of the related terms and that, accordingly, relations are ultimate concepts as basic as predicates. However, curiously Russell employs the recourse of transforming predication into a relation, rather than completely denying the subject-predicate pattern. Thus, instead of resting only on the irreducibility of asymmetrical relations (as he did in FIAM and previous works), he also adds that there is no point in intending to reduce relations to properties (like Bradley intended, according to Russell) because properties are already relations. However, the traditional argument also appears with regard to order in general: in '*A* is before *B*' we do not imply *predicates* of *A* or *B*, since position cannot be a predicate (difference in position being reciprocal and not transitive) but, simply, nothing different from the point itself (the same happens with greater/less, etc.).

Russell is doubtlessly under the influence of Moore's relational theory of judgment, but we should not forget that, although Russell classifies relations, the classification itself depends, like in Kant, on the various kinds of propositions (or judgments): 'The classification of relations is, therefore, the classification of the types of propositions. This fact brings to light the importance of a classification of relations, and the relation of such a classification to Kant's deduction of the categories'. We have, then, another example of 'presupposition' (or analysis) which tries to discover the true meanings in our language on relations. This will be the link with FIAM and POM1, where Russell was still a Kantian.

The article ends with four conclusions and a problem. The conclusions are: (i) there are no symmetrical relations; (ii) there is only one reciprocal relation: diversity; (iii) there are four types of relations (the ones mentioned in the classification) that are sufficient for mathematics; (iv) it is doubtful that there are subject-predicate propositions: they are reduced to inclusion or to the type of one-sided relations. The problem is Bradley's famous objection against relations: if we admit them, we have also to admit relations between each relation and the related terms, which leads us to an endless regress. The final lines of the article are devoted just to say that the solution of such difficulty would be the most valuable contribution by a modern philosopher (however, as I show in my *1990d*, Russell never found this solution).

We now come to POM1. Relations impregnate the entire content of the work, where we even find a whole chapter on them. Russell starts from a consideration of order that reduces it to asymmetry and the notion of 'between' (POM1, IV/1): 'what is meant by order is a certain type of relation between three terms, such that one of them is asymmetrical with respect to the other two'. In this case, terms are 'positions' which are related to terms like magnitudes are related to quantities. In the case that some of them have symmetrical relations, we will have to reduce them to 'sameness of position', in the same way that magnitude was reduced to 'sameness of relation'. Thus, position comes to light as a fundamental conception. There is no asymmetrical relation common to all positions, but all of them are characterized by a difference of sense (IV/2). The 'distinction of sign' is a quantity expressing the fact that asymmetrical relations have two senses from which the difference of adjectives of two related terms is made possible. That is why quantities with sign depend on order and asymmetrical relations, which 'are of absolutely vital importance for a sound philosophy of Mathematics, and it is they that best exhibit the inadequacy of the traditional logic, according to which every proposition, at bottom, is one assigning a predicate to a subject' (*ibidem*).

The proof of the impossibility of reducing asymmetry to the predicative pattern (POM1, IV/3) was already considered in section 2.5, where Russell deduced the contradiction of relativity by assuming the traditional theory as an hypothesis. We saw also how external relations are the only possible resort to avoiding the contradiction: 'we cannot hope, therefore, so long as we adhere to the view that no relation can be "purely external" to obtain anything like a satisfactory philosophy of mathematics'. There are several things to say about all this.

Firstly, it is a pity that the classification which Russell implicitly starts from is lost (the POM1 manuscripts lacks f. 34). But I do not think that it was the same one as that of *m1899a* (which we saw above), since Russell insists that the contradiction only appears in the relations of the so called 'fourth' type (characterized by asymmetry and illustrated by causality, spatial and temporal position, and inequality), whereas in the former classification he considered *two* types of asymmetrical relations, causality being then referred to the first of them (transitive relations), and inequality being referred to reciprocal relations (those being symmetrical and non-transitive). In any case, the truth is that asymmetrical property has now the pre-eminence.

Secondly, the reducibility of symmetrical relations (which depends on 'identity of content') is confirmed through the principle of abstraction (which is not yet explicitly mentioned), until turning them into 'sameness of relation' to a term that represents the common property: 'the so-called properties of a term are, in truth, other terms to which both have the same relation'. Symmetrical relations are thus discarded as genuine relations.

But, thirdly, in retaining asymmetrical relations as irreducible an old problem again appears: that of a possible underlying form of predication through the relational theory of judgment:

> All irreducible relations are unsymmetrical,[1] and there must be irreducible relations, since a proposition must contain two terms at last, and the proposition constitutes a relation between them. The distinction of things and qualities, which depends upon the doctrine of subject and predicate vanishes.

---

[1] It seems to be the only place where Russell uses this term at this stage.

In this crucial passage, Russell allows us to discover his deepest thoughts, which continue to be indebted to a particular logical and ontological analysis of the proposition, and besides, does not seem compatible with his doctrine (sometimes formulated in these manuscripts) that the proposition has to show a unity as being irreducible to its terms (though of course relations are also terms). As we saw before, the pre-eminence of relations over the subject-predicate pattern is carried out simply by introducing in this pattern the strategic modification of placing both concepts at the same level, by turning them into logical subjects (like Moore), though maintaining the resulting general schema as a paradigm. In this way we obtain the appearance of having overcome the predicative point of view, namely the distinction between things and qualities. However, on the other hand, we also are dangerously near the relapse into Bradley's belief that the subject of all propositions is the same: Reality; though for Russell this 'Reality' consists of the same logical-ontological material as propositions: the later named *logical atoms*, which are already implicit here. I think this is a consequence of the subtle difference between monism and monadism that Russell intended to maintain, not always successfully. In any case, only on this ground can Russell's philosophy at this stage be understood, especially his theory of definition, the non-distinctions between some of his main doctrines before and after the Paris Congress of 1900, and, finally, the essential continuity of his philosophical method.

2.7. The method of definition

The main goal in AMR is clearly stated in the introduction and the first chapter: to discover the conceptions and necessarily presupposed judgments in pure mathematics. This forced Russell to pose the basic problem of indefinables (needed for avoiding a vicious circle) and to relate them to intuition: 'In order that it may be possible to use a conception thus left undefined, the conception must carry an unanalyzable and intuitively apprehended meaning'. That is why, once such conceptions are discovered, we must not expect that 'they should be defined or deduced', but only that 'they should be merely indicated'.

Thus, the first problem is brought up: on the one hand, the method of discovering indefinables is presupposition; on the other, only intuition makes the access to them possible. The explanation of the problem lies in Russell's progressive evolution from FG. In that work he was a Kantian and he thought that the axioms to be discovered are presupposed, in the sense that they are accessible to intuition (in the Kantian sense). However, starting from Moore's influence, he tries to turn this intuition into a purely logical process, making it synonymous to 'logical priority', which would objectively exists between logical subjects (terms) and also between propositions (synthetic relations between terms). However, Russell still thinks (in AMR) that, since indefinables and indemonstrables are intuitively 'discovered' (they are a priori), the process has to be reversed by *deducing* the judgments of mathematics from those axioms (it is the twofold process of FG; see 1.5 above).

AMR is an unfinished work, but according to the preserved plan it had to go from number (an indefinable) to infinitesimal calculus, space, geometry and 'abstract' physics; always through the analytic method in searching for the presupposed indefinables and indemonstrables. However, as I pointed out before, Russell thought that the only guarantee of coincidence between the processes of presupposition and intuition was to prove, later, that the order could be reversed until deducing everything from the discovered principles. It would be a procedure similar to that from FG: 'When the analysis of mathematical methods is ended, I shall return to more fundamental questions, and endeavour to deduce the mathematical methods from a priori principles, instead of adopting, as here, the converse analytical order of investigation'[1] (III/5; they are the final words of the manuscript).

In FIAM there are only some scattered remarks more or less relevant to the method of definition, but only one of them deals directly with it, though it is very useful to us as it states the continuity between AMR and POM1. It is a passage proceeding from the material on magnitude and quantity. In short, it holds the need for choosing between the two possible theories by trying to discover the true indefinables (simple) concepts, and by arguing that it would be a *serious philosophical mistake to regard definition as conventional* (FIAM, 9):

> Every concept is necessarily either simple or complex, and it is not our power to alter its nature in this respect. If it is complex, it should be analyzed and defined; if simple, it should be used in defining other terms, without itself receiving a definition. Thus equality either may be analyzed into sameness of magnitude, or it may be not so analyzed. If our former theory was correct, the present theory must be incorrect. It does not lie with us to choose what terms are to be indefinable, on the contrary, it is the business of philosophy to discover those terms.

In this way it is clearly established for the first time, in Russell's view, what has to be the basic method of any philosophy. Nevertheless, Russell does not offer any criterion apart from mere intuition (not even conceptual economy, which is a later device). It is true that through an incipient 'principle of abstraction' certain relations and properties can be dissolved, but the argument adduced to preserve magnitudes as indefinables is that, since they are essentially unequal, the main problem ('to see why some quantities should be equal and other unequal, unless there is some respect in which those that are unequal differ') does not arise. However (as is recognized) this argument depends on the *meaning* of the terms: 'when we consider what we mean when we say that two quantities are equal, it seems preposterous to maintain that they have no common property not shared by unequal quantities', but, how do we know what terms really have a

---

[1] In the later attempts this twofold process is not even projected, in so far as Moore was reaching the pre-eminence over Kant (as I show in my *1990a*).

meaning without resorting to intuition? Russell here seems to elude Moore's criterion of ordinary language, doubtless because of his great interest in constructing a chain of definitions.

POM1 is the point of confluence of the previous trends, which are framed into a global context where logic, ontology, analysis, plurality and definability are related to each other, drawing the general method with which Russell would face the Paris Congress. Firstly the identification between what is indefinable and what is purely logical becomes confirmed. Numbers, for instance, are presented as indefinable, not only because they *cannot* be defined by addition or through collections, but mainly because they do not imply anything from which they can be asserted. They are pure and simple concepts; if they implied something, this would be the whole set of its constituents, which would make them complex. The impossibility of finding a definition turns into a negative reason: 'there is therefore no reason to deny that all numbers are indefinable and simple' (I/2).

In any case the capacity to imply is brought up as a criterion also applicable to propositions with the same goal: to evaluate its simplicity (POM1, IV/4). The propositions implied have to be considered as the *parts* of the proposition that implies them (except for mutual implication). However, the criterion is not purely logical since there is no secure (algorithmic) method to find the presuppositions: we have only the immediate intuition of what is simple; but the knowledge of a proposition does not imply (epistemologically) the knowledge of its premises. To think so is to ignore the difficulties of philosophy: 'For these difficulties are almost entirely concerned with the discovery of premises involved in propositions known to the plain man' (VI/2).

The references to Kant are limited to accepting the synthetic character of arithmetical propositions and to the effort of establishing a nexus between what is synthetic and what is presupposed. Thus, in $7 + 5 = 12$, the sum is not a part of the premises, but it contains a new simple concept that is presupposed in them (POM1, I/3). It was doubtlessly a comfortable position for Moore's 'intuitionism', but it made the acceptance of Cantor impossible. Russell considers the arithmetization of mathematics (and its withdrawal from intuition) as an attack against Kant, and one much stronger than that proceeding from non-Euclidean geometries (V/1): this process has made 'that the continuity of space and time, however it differs from that of numbers, does not differ in a way which is relevant to the Calculus'. As we shall see below, here the distinction between what is technical and what is philosophical is already present, which would be the main defence against Cantor.

This view leads to a pre-eminence of logic, which allows Russell to regard philosophical analysis as a way to bring to light that which is genuinely real in any field of science, as it occurs in mathematics: 'For a philosophical analysis of mathematical ideas, it is of the highest importance first to divest them of that clothing in artifices which is required by the purposes of calculation, but is liable to hide, only too well, the inner nature of the objects of the science' (POM1, IV/5). This essentialism is, then, inseparable from the Boole-Peirce-Whitehead tradition and it leads in a natural way to the pre-eminence of logic with just one more step. This step was taken by Russell starting from the Congress of 1900 and it allowed him to clarify which of the three viewpoints involved (logic, arithmetic, ontology) would have the primacy.

Accordingly, *definability*[1] shows the essence of the concepts through the constructive explicitation of their components: 'definition can only consist in enumeration of indefinable constituents'. The only problem will be that of complexes being genuine unities, where enumeration cannot be admitted:

> A complex unity cannot be defined as having no presupposition, but only as differing from all its presuppositions... For complex units are essentially propositions, and in these, as we have seen, the concept which is not a term, whether omitted or treated as a term, will not yield the same result as in conjunction with the concepts which are terms.

But if 'one' is equivalent to 'indefinable', then *undefinability itself is indefinable*, and the only operative solution is to resort to the ontological concept of plurality through its identification to the concept of definability: 'If unity means undefinability it seems to follow that plurality means definability. Since plurality is one, plurality or definability is indefinable. Since the definable is logically subsequent to the indefinable, the plural is logically subsequent to the singular'. In this way the pluralistic concept of definition is definitively established.

Moore's odd philosophy is then the one that, through a fusion of the logical, ontological and epistemological viewpoints, provides the context needed for framing Russell's attempts to constitute a pluralistic analysis of mathematics. Moreover, as there are interrelated complexes, analysis is impossible and their meanings are not the sum of the meanings of their parts. However, such entities are only complexes from the viewpoint of the meaning. When Russell found techniques powerful enough (as the theory of descriptions), he even analyzed some of these unities by dissolving them into genuine structures, but then a certain ontological simplification had already taken place, which was not yet assumable in the present framework of such an ontological exuberance (which prevailed in POM). In any case there is a thing completely sure at the end of the manuscripts: the identification of pluralism, extensionality and definability. Once again: 'pluralism seems involved in the very foundations of all mathematical reasoning, as appears from the fact that the extension of concepts is essential'.[2] This pluralistic notion of definability, though still too close to an extralogical intuition (if talking about a 'logical' intuition makes any sense), would be the one that will guide Russell's method in the next stages.

---

[1] All that follows, unless otherwise indicated, belongs to the chapter entitled 'Plurality' that appears at the end of the POM1 materials.

[2] This passage proceeds from some pages on 'manifolds' belonging to the POM1 materials, and the former one from the 'Plurality' chapter (see the former footnote).

2.8. The gradual approach to Cantor

The object of this section is to explain something still unexplained: Russell's transition from a strong rejection of Cantor's transfinites in the publications and manuscripts before 1901, to an almost enthusiastic acceptance in POM. Grattan-Guinness[1] has attributed all the merit of Russell's acceptance of Cantor to the Paris Congress; however, even though the Congress accelerated this acceptance, an accurate study of the relevant manuscripts (especially those from 1899 to 1900) shows how this acceptance was preceded by a gradual approach, which was framed into a context already evolving more and more toward an incipient 'logicism', characterized by the pre-eminence of the 'logical' point of view in the search of indefinables.

Therefore, the required ground for accepting Cantor was already stated, in so far as the progressive approach and penetration in the problems of infinity and continuity were one of the catalysts of Russell's evolution at this crucial time (along with a probable gradual better understanding of Cantor's theory in itself). This is also related to Russell's acceptance of Peano's logical methods, but not only as a filter through which he could assimilate Cantor (Grattan-Guinness *1980c*, 63), but rather as a set of techniques that, among others, forced him to determine with precision a ground for definitions and their philosophical import. Just a glance at the manuscripts shows that as *before* as *after* the Congress, Russell accepted some of Cantor's ideas from the mathematical viewpoint (in order to explain the technical essence of continuity and the infinite). The 'philosophical' value of Cantor's constructions was the only thing that really evolved (as Garciadiego *1985b* was, to my knowledge, the first to recognize).

2.8.1. The first contacts and opinions

In a letter to Jourdain of 1917 (in Grattan-Guinness *1977a*, 143-4) and as a response to a question about the source of his interest in mathematical logic, Russell wrote that he knew Cantor through Hannequin *1895a*, together with Couturat *1896a* (curiously, both works provided opposite approaches to Cantor). Russell adds: 'I read all the articles in the *Acta Mathematica* carefully in 1898, and also the *Mannigfaltigkeitslehre*. At the time I did not altogether follow Cantor's arguments, and I thought he had failed to prove some of his points. I did not read the articles of 1895 and 1897 until a good deal later'. Finally, he writes that it was Peano who led him to a logical treatment of arithmetic.

---

[1] His *1980c* about Cantor's influence on Russell was, to my knowledge, the first attempt to globaly throw some light on this important point.

In spite of some incongruities with the rest of the available information,[1] the passage is interesting and can be complemented by another (AB1, 127) where he explains how in the beginning of 1896 'I spent the time reading Georg Cantor, and copying out the gist of him into a notebook'. Russell immediately adds the following judgment: 'At that time I falsely supposed all his arguments to be fallacious, but I nevertheless went through them all in the minutest detail. This stood me in good stead when later on I discovered that all the fallacies were mine'. Russell leaves undetermined the moment when he realized Cantor was right,[2] but as we shall see in what follows it was a truly long and winding process.

From Russell's view, the deepening of his knowledge on the foundations of mathematics was related to the abandonment of the old idealistic and Hegelian doctrines. As he explained later, such an abandonment allowed him to regard mathematics as something true, 'and not merely a stage in dialectic' (*1944a*, 12). Russell seems here to simplify things a little, but the truth is that historically his abandonment of Hegelianism was followed by a total devotion to the founding of mathematics. His evolution on Cantor shows that the idealistic criticism against mathematical constructions survived to the official abandonment of idealism,[3] especially because of Kant's philosophy of arithmetic, which seemed to be the only acceptable one, at least as the only serious alternative to empiricism (*1944a*, 1). In the publications we find, however, a complete rejection of Cantor. In *1896d* Russell follows Hannequin's criticisms in rejecting any possibility of applying the number, which is 'discrete', to continuum:[4]

> the attempts of Cantor to extend the conception of pure number so as to cover continua... seem to me, ingenious as they are, to be open to even severer strictures. For Cantor's second class of numbers, by which he hopes to exhaust continua, begins with the first number larger than any of the first class; but as the first class (the ordinary natural numbers) has no upper limit, it is hard to see how the second class is ever to begin.

Russell's position becomes more moderate in *1897b*, a review of Couturat *1896a*. The change consisted especially in the partial acceptance of Couturat's detailed arguments *concerning*

---

[1] Russell frequently suffered some confusion when mentioning dates and other similar information. Grattan-Guinness (*1980c*, 63) accepts 1898 as the correct date, according to the evidence of the quoted passage, but that from AB1 (which I also quote), i.e. 1896, seems more reliable to locate the first reading (for a full study of that point, see my *1987a*, 384 , note 49).

[2] The 'popular' version of the conversion takes place in PRM.

[3] Russell always insisted that the new mathematics was chiefly a clarification of the confusing traditional view.

[4] *1896d*, 412. Spadoni (*1977a*, 155 ff) quotes an unpublished essay of June 1896 ('On some difficulties of continuous quantity') which illustrates Russell's opposition to the analysis of continuum: 'even the humblest philosopher, one would think, must grow indignant at Cantor's ω'. Spadoni also quotes the following passage by Bosanquet: 'Being one-sided, the idea of infinite number is self-contradictory ... It follows from this that infinite number is unreal'.

mathematics, where Cantor's usefulness and unquestionability 'is by this time almost self-evident'. Nevertheless, he is 'philosophically' rejected through a series of arguments which are somewhat hard to determine with precision. The ground to all of them seems to be a general objection: the admission of transfinites leads to contradictions even in its process of formation (see 1.2 above). Starting from the standard conception that any number has to be a completed and defined whole (a Kantian synthesis), Russell finds it unacceptable that Couturat 'boldly' (*1897b*, 115) regards a series as given, as soon as a law permitting to obtain all its members is available (although he also admits that it was the only possible defence of Cantor).

Russell only adds one more argument: the infinite is a deduction drawn from the premise according to which space and time are not mere relations, but something else: 'in all the cases where infinity is unavoidable, there has been some undue hypostasising of relations, which makes the attainment of a completed substantive whole impossible' (*1897b*, 119). Fortunately we have FG to try to clarify so cryptic a passage. There Russell holds (FG, 195-6), in the context of his relational theory, that space is conceptual (a mere order), and that when its relations are hypostatised the whole collection of them appears as contained in *empty space*. However this empty space is nothing but the logical possibility of such relations. When it is regarded as a genuine entity it appears as unnecessary because the set of relations is sufficient to give an account of everything necessary to geometry (especially when an unextended *matter* is later introduced to avoid contradictions; see 1.5 above); but the antinomies arise only with empty space, whereas 'when space is regarded, so far as it is valid, as only spatial order, unbounded extension and infinite divisibility both disappear' (FG, 196).

Curiously Russell rejected in PL this view of space (already in Leibniz), to hold absolute space because of metaphysical (pluralistic) reasons, as the only one capable of embracing at the same time external relations and the immutability of concepts. However PL served Russell, not only as an attempt to destroy Leibniz's space, but also to adopt a somewhat different view as regards infinity. At this time he thought (as it can be seen in the manuscripts) that the transfinites must be rejected,[1] though the actual infinite must be admitted (which is nothing but a repetition of Leibniz's position; see Ishiguro *1972a*, 140). There is still other allusion in PL: that recognizing the principle according to which infinite aggregates have no assignable number, regarding it as 'perhaps one of the best ways of escaping from the antinomy of infinite number' (PL, 117), which was just the view adopted in the manuscripts to elude the acceptance of Cantor.[2]

On the other hand, in the unpublished writings the rejection is not so strong. AMR shows no interesting references to infinity or continuity that are not framed into the general context of the 'contradiction of relativity' (see 2.5 above). However neither Cantor nor any of his theories are

---

[1] In the 1937 preface to PL we read that Russell then knew 'little of mathematical logic, or of Georg Cantor's theory of infinite numbers' (p. viii), to which he attributes the description that he makes there of mathematical propositions as synthetic.

[2] In the article read to the Congress (like in *1899b*) he rejects that the concepts of order and correspondence can *create* series (*1900b*, 706).

mentioned, despite that he obviously knew them (as we have seen). Perhaps the firm rejection permitted him to still ignore the important theory.

In the general plan to FIAM none of the eight parts is devoted to Cantor, though in the actually written material there is a definition of continuity ('IV. Continuity, Zero and Infinity'; a loose chapter perhaps belonging to POM1) in order to give a precise meaning to this term. This is already a sign of the influence of Dedekind and Cantor, who had similar motivations. Russell's definition is: 'Continuity applies to series (and only to series) whenever these are such that there is a term between any two given terms', i.e. anything not being a series will necessarily be *discrete*. The intuitive character of the definition has to be emphasized, which coincides with Russell's belief that the philosophically admissible definitions should reduce the defined concepts to immediately known constituents. However Russell later added a note (perhaps on rereading Cantor) in which he recognizes the 'technical' superiority of the German mathematician:

> the objection to this definition is that it does not give a fixed property of a collection, but depends sometimes upon the ordering of the terms, e.g. the rational numbers, in order of magnitude, fulfil the condition, but the logical order do not fulfil it. This objection does not apply to Cantor's definition.

The note does not develop the mentioned advantage for Cantor's definition, but it is already a sign of the gradual approach at this time.

What distinguishes POM1 as for its partial acceptance of Cantor[1] is that this work devotes *an entire part* (nine chapters) to 'Continuity and infinity'. Russell's general view consists in technically admitting some of Cantor's results and philosophically rejecting transfinites as contradictory, offering an alternative based upon the notion of infinite divisibility. Here we shall consider only the relevant arguments, which will have to be explained (justified) at a further stage.

According to Russell's general viewpoint (V/I) it is necessary to accept the trend of 'modern' mathematics that infinitesimal calculus does not require the resort to 'intuition', but the study of the series in general, without relations to the special peculiarities of space and time. This coincided with Leibniz rather than with Newton, and especially with the progress of the theory of numbers, which constitutes, according to Russell, the true attack against the doctrine of Kant's a priori intuition (and not that coming from non-Euclidean geometries, which can still be accepted from Kant's point of view, as Russell 'proved' in FG). In the chapter devoted to continuity (V/3) Russell starts from the distinction between Cantor's analysis and that which he himself offered in the third part of the work (today lost, but perhaps similar to that of FIAM). He describes with an apparent approval Cantor's attempt to carry out the conceptual analysis required for a true

---

[1] In VI/1 it is admitted that it is impossible to define continuity without reference to order (by quoting Couturat *1900c*). With this we verify he knew, at least in an indirect way, part of Cantor *1895a* (since Couturat *1900c* rested on this work).

understanding of this notion. To continue he offers an explanation (based on Cantor *1883a*[1]) in terms of *perfect* and *well-connected* series, adding some comments on the notion of power and some examples. The important thing for us is to notice how, after pointing out the relation between continuity and infinity (and transfinite numbers), Russell expresses his valuation in the well-known terms of the distinction between mathematics and philosophy. For him what is technically valuable can be philosophically unacceptable: 'I wish to examine what use, if any, philosophy can make of this mathematically invaluable analysis of continuity'. However, in the conclusion to the chapter Russell discards Cantor's continuity by reducing it to his own definition of this concept.

Chapter V/5 is the key to understanding Russell's position about Cantor, since it is devoted to the transfinites. At the beginning we find a global evaluation that sums up a mixture of admiration and rejection. On the one hand he admits that the mathematical theory of infinity began with Cantor, who abandoned the cowardly policy of eluding the concept, taking the skeleton out of the cupboard,[2] in order to deny it really was a skeleton, and 'established a branch of mathematics logically prior to the Calculus and even to irrationals'. On the other hand, he points out the incapacity of Cantor's construction for solving the true problems of infinity and its antinomies:

> I cannot persuade myself that his theory solves any of the philosophical difficulties of infinity, or renders the antinomy of infinite number one whit less formidable. Like most mathematical ideas on the subject it consists of a skillful combination of the two sides of the antinomy in the proportions most useful for obtaining results.

From here on Russell offers some particular arguments, of which we are going to follow only what is truly essential.

Against the validity of Cantor's 'second principle of generation', leading to ω (see 1.2 above), i.e. against the existence of any number not obtained through successive additions, Russell argues that it violates the universal principles of arithmetic: there is no link between the series of naturals, which has no last term, and the series that begins with ω; the link is only imaginary. Besides, we cannot know that ω comes after all finite numbers unless it belongs to the same series as they do; but, if ω is assimilated to a limit, it cannot then be defined by means of the series (the limit should be independent of the series). Finally, Russell denies that ω possesses uniqueness adducing that it is equal to the number of even numbers, prime numbers, etc.

Concerning the *true infinite* Russell analyzes various possibilities coming to the conclusion that it is necessary to reject the existence of infinite numbers. During the discussion he mentions an axiom (from the lost part III) according to which a given collection of many terms should contain

---

[1] In spite of quoting Couturat *1900c*, where Cantor's two definitions are described (perhaps the article came to him when he had already composed the text).

[2] Chapter 37 of POM begins with exactly the same words; but after the first paragraph we already read some enthusiastic praises of Cantor that are still lacking here.

some particular number of terms. According to Russell, the acceptance of this axiom leads to the belief that there has to be a definite number of numbers, but if we deny the axiom, the only possible ground for transfinites vanishes. Russell recognizes that the parts of a whole should have *a* number, but he rejects the axiom because it leads to contradictions, whereas by denying it we are led only to 'oddities'. The final balance is: 'In any case, the difficulties on infinity are not obviated, but are only dragged into light, by the transfinite numbers; and this is, for philosophy, their principal and invaluable merit'. Now can we try to give an explanation of the reasons for this rejection.

2.8.2. The reasons for the rejection

I am convinced that the *philosophical* rejection of the new mathematical constructions must be explained through an entire global and coherent conception rather than through the detailed discussion of this or that particular argument. With this I am not suggesting a psychological reply to the problem, in terms of which Russell's arguments were *only* a curtain for the true 'motives' (although perhaps something of this kind happened). I think that one must attempt the construction of a conceptual framework where these arguments are inserted as inferences carried out from a *genuine philosophical rejection*, whose details are relatively irrelevant. I also think that this conceptual framework already contained the seed of a subsequent acceptance. Only in this way we could explain that Russell's acceptance of Cantor (like that of Peano) would play the role of a catalyst with regard to an independent project already previously established. As always, Russell regarded it as vital to be able to continue the work by applying the same analytic method based on constructions-definitions. In what follows I shall try then to *interpret* this global rejection of Cantor, resting on the ideas pointed out above (and on other materials from my *1897a* which I have no place to develop here).

The reasons for the rejection during the Hegelian 'stage' are difficult to establish with precision, but fortunately they are less important given that the Hegelian inspirations finished very soon. However, they have the strange property of partially coinciding with the subsequent reasons. Russell doubtlessly was then forced to reject anything that would be presented as a literal attempt to give an account of such a complex notion as continuity. In pointing out what is discrete as the only rational possibility for our understanding, he was defending the neo-Hegelian doctrine according to which we can only arrive at the 'true' continuity after a long process of dialectic ascents, but never through a simple reordering of the available data, no matter how skillful this was. This Hegelian belief also led Russell even to deny the 'technical' possibility of constructing infinity through 'logical' means: the true logic has to be identified with reason, which can overcome the antinomies, rather than with understanding, which is merely algorithmic.

The reasons proceeding from Kant should have played a certain role as well, especially in the philosophy of arithmetic. Russell possibly saw with skepticism Couturat's criticism of Kant's antinomies, for though they were also non-genuine for him (they could be overcome in the Hegelian standard way), they cannot be 'solved' through the elimination of one of the two sides (the 'mistaken' one). Kant thought that the quantitative syntheses are necessarily successive and are constructed through addition of parts; but for Cantor infinity is more a whole than a sum of parts: it has to be necessarily constructed by resorting to a law or principle of generation that gives an account of the internal order of its terms taken as a whole. The mechanism of generation will automatically yield the infinite class with only stating the law. At the same time, this viewpoint presupposes the existing opposition between Kant's philosophy of arithmetic and Cantor's infinity: for Kant any number has to be the result of successive syntheses (a view accepted by Russell[1]), whereas transfinite numbers cannot ever be constructed in this way. Here the opposition is again based on intuition, which has to be the only foundation of arithmetic according to Kant.

However Russell could not accept Kantian intuition, for it intended to be capable of founding the empty space as a truly 'objective' entity, which was infinitely given at the same time. But from the point of view of his logic at this stage, it (mistakenly) supposed to regard a mere set of relations as an entity. Hence his argument that infinity, as a conclusion, depends on the *premise* constituted by a substantial (non-relational) space; and this premise, again, depends on regarding the relations constituting space as an entity (in spite of the fact that space is only the mere logical possibility of those relations). However, following Bradley, these relations (together with the related terms or qualities) are not true entities that can subsist by themselves; therefore we can arrive at the (unjustified) inference of an empty space only through an hypostatising of relations. As Russell denies the premise, that he regards the only possible ground for the empty space, he also denies the conclusion. Consequently, if space is real (i.e. if relations are real), then infinity must be admitted.

It is easily imagined that the next step, to admit the antecedent and to deny the consequent, was not very pleasant. However Moore's influence led Russell to admit the full reality of relations (and that of their terms), which quickly led him to absolute space. (He even admitted —in PL— that 'the monists' need to deny the reality of space, since they depend on the predicative pattern and the notion of substance, with which his own previous view was obviously affected.) That is why the strategy used to reject Cantor had to be changed. However this change did not force him to dispense with some arguments, like that of the synthetic character of the arithmetical propositions. We have already seen how it was a literally indispensable argument, which besides seemed to coincide to some extent with the relational theory of judgment (which regards all propositions as synthetic, though in a somewhat different sense).

In this context we must mention another recourse, which had already been used to reject certain constructions in geometry (see 1.5 above): the distinction between mathematical and philosophical definitions. In short, philosophical definitions have to give an account of the essence of the defined entities, whereas mathematical definitions can be limited to offering sufficient and/or necessary properties. Therefore the first ones have to be based on the criterion of the *intuition* of the component elements of the *definiens*. This old theory coincided very well with the method that Russell had learned from Moore: a definition has to be an analysis into indefinable simples previously known through the immediate intuition, starting from ordinary language as

---

[1] The definition appears in *1897c*, *1896c* and other unpublished former writings.

raw material. However, Cantor's fundamental achievements supposed a strong attack against some features of the new method.

Cantor's view coincided with Moore's method since he tried to find the *essence* of certain concepts (continuity, etc.); it was, then, acceptable. However, instead of using the criterion of ordinary language (i.e. to start from the usual senses of the terms in order to obtain the 'true' meaning), he intended to construct the true meanings by means of previous defined concepts (order, transfinites, etc.). I shall refer to them below, but here I would like to say that this reduction violated the requirement that the materials employed in the constructions were immediately known by intuition: a complete reduction would eliminate the entity to be defined. It was consequently impossible from Russell's point of view to avoid the recourse to his distinction between mathematical and philosophical precision: we can put to work the defined concepts by framing them into a mathematical context of operative relations to other concepts, but they have to maintain their own indestructible philosophical essence.

As for transfinites, it is clear that the most serious problem was their contradictory character. However all the contradictions they involved could be reduced to the difficulty of obtaining the first term of the second class of numbers without resorting to addition. At this point Russell saw an insuperable gap, in the same way as he saw it concerning Dedekind's *cuts* (see 1.2 above). However the true reason for rejecting the construction seems to be that, on breaking with the usual intuitiveness, it requires a certain recourse to our capacity to *imagine*, in order to embrace the way in which the second law of generation manages to make an entity equal to the set of all the numbers satisfying it. This 'mental effort', that is already found implicit in Cantor's definition of set (through his recourse to the 'whole' unifying the elements), also resorts to infinity: only at infinity can the second class of numbers 'begin'. We must remember here that for Russell the recourse to the *unity* of an infinite whole had certain 'psychological' connotations (as can be seen in his attacks against Leibniz because of similar reasons). Cantor improved his first theory very much by resorting to order types in his *1895a* (see 1.2 above), but Russell knew this work only after the Paris Congress of 1900, so that he only worked on the first version.[1]

All that leads us once again to the concept of intuition. Here we must distinguish between the 'imaginative' type (required in Cantor *1883a*, as we have seen) and the 'logical' type, through which we arrive (according to Moore and Russell) at the logical subjects (the pure and simple concepts). The distinction can be the root of another of the true reasons for the rejection. The interchangeable (and lacking any particular nature) unities that form the elements of Cantor's set theory seemed to be well adapted to Moore's ontology (in both places we have: indivisible = indefinable); they contained even the seed of the subsequent foundation of all mathematics (as Russell attempted from 1900 on). However, Russell probably did not like the empirical character of these elements that, according to Cantor, had to be *abstracted* by us as an 'intellectual representation' of the external reality (of nature). On the other hand, the 'logical' intuition referred to by Russell is equivalent to an immediate contact between the mind and the concept (a 'true meaning'). If we do not take into account this requirement our definition will be limited to giving

---

[1] However through Couturat *1900c* he knew about the improved version *before* the Congress of 1900.

us only some feature of that concept or, at most, to state some relation to another entity, but without bringing *its essence* to light. This is the reason why Russell wrote things as the following: 'the definition does not define in the philosophical but only in the mathematical sense: it gives a mark or criterion, but presupposes the notion of which it is a mark' (POM1, V/9). Of course, the kernel of the problem lies in the ontology. Both Cantor and Russell were anti-formalist and Platonic, but in different ways (see Grattan-Guinness *1980c*). To reduce the differences to only one formula, we can say that, for Russell, Cantor was too formalist but not Platonic enough.

Cantor defined infinity (like Bolzano) through the correspondence between a set and one of its parts. In doing so he added the objectivity of the transfinite numbers to that 'formal' property, showing how these new numbers fulfil the usual properties of numbers. This criterion, that was also formal since it was limited to exhibit the consistency (or absence of contradiction) of the new construction,[1] should however have seemed still insufficient to Russell (like to Frege). However Cantor was also a Platonist when he insisted that these entities were objective: they depended on the objective existence of things in nature. With that he intended to overcome formalism and arrive at the 'intrinsic' nature of the defined entities; in this way, they were not limited to being mere signs fixing properties or mere formal 'symbolic' constructions. However, again, his Platonism was a *sui generis* feature; when he referred to the nature itself of the things he alluded (we saw it before) to the intellectual abstractions or 'reflections' of 'external' processes. Thus, his definitions of 'set' and 'power', as well as his guarantee of the existence of infinity, proceed from the abstraction of phenomena existing in nature, which have to be the starting-point giving objectivity to the *foundations* of the constructions. Once these are secured, mathematics is completely *free* of building entities with only offering a guarantee of consistency. That is why Cantor also rejected, like Russell, the axiomatic point of view (see Grattan-Guinness *1980c*), i.e. the possibility of presenting indefinables through only the axioms exhibiting their relations.

On the other hand, for Russell the mind is not 'free': it is limited by its 'vision' of the logical absolute contents and their intrinsic relations. Although Cantor called himself a mere 'secretary' when describing the objectivity of what he 'saw' (i.e. without inventing anything), he was not referring to the same idea as Russell. For Russell it was not possible that the objectivity of the concepts was shown only through these or those relations between them; this would be formalism. The problem was precisely the admission itself of those concepts, which have to be 'seen' before they were operatively related in some way. Therefore, although both rejected 'arbitrariness', the essence that they intended to show were of a different kind. Cantor's strange mixture of formalism and Platonism had to be rejected according to Russell, although he himself would soon fall into a similar situation on intending *at the same time* to present his constructions as 'logical possibilities' and as genuine 'essential entities'.

One final remark: for Cantor the construction of transfinites and continuity was based upon order, and that, again, proceeded from a much wider conception: set theory. The power of this theory shook even the supposed irreducibility of number. The entire work by Cantor can be regarded as a true logical analysis of this notion. He even was convinced of the need for logically

---

[1] As Grattan-Guinness points out (*1980c*, 84) it was a pre-Hilbertian formalism.

founding it. On the other hand, for Russell one of the basic dogmas before the Paris Congress was the *undefinability* of number. Had he admitted that number was definable, then another indefinable point of departure for the whole constructions of mathematics would be needed; that was what Moore's logic and ontology required.[1]

At this point the real link between Russell's acceptance of Peano and that of Cantor appears. It is obvious that both took place at the same time and in a common context. Peano was accepted to be able to continue with the process of reductive definition. This supposed to concede a certain conventionality to the 'primitives'. Therefore, if the process of definition does not have to involve *absolute* indefinables, then other conventional constructions could also be accepted. Thus, though Cantor's transfinites did not guarantee wholly intuitive starting-points (in the logical sense), at least they had the virtue of submitting them to an objective order of relations that could be defined with precision. In both cases the requirement of a total intuitiveness for the 'simples' had to be moderated, or at least be reduced to the 'philosophical' definitions. Now we can see the kernel of the problem: if mathematics is reduced to philosophy, and logic is the foundation of philosophy (as required by Moore's logico-ontological parallelism), then the distinction between mathematical and philosophical definitions is no longer needed. The fact that in POM the distinction was maintained is already a sign of the partial incompatibility between Moore's and Peano's 'logicisms'.

---

[1] For the same reason Dedekind was also rejectable. From his writings the reduction of mathematics to logic could be inferred, which supposed the same analysis of the concept of number through a number of axioms that 'implicitly' define it.

3. The contribution of Peano and his school

The main object here will be to evaluate the influence of Peano and his school on Russell, starting from the contact that began in the Paris Congress of 1900. This means delving deep into Peano's own work (and of some of his followers), considering it in itself as the object of study. In this way we shall set the ground to be able to determine with precision what Russell took from Peano, once Russell's relevant works are studied in the next chapter. The second object will be to criticize some commonplaces about Peano, whose works are quite unknown,[1] perhaps for having been eclipsed by the works by Russell and Frege. Some of these commonplaces are the following: (i) Peano's main achievement was to invent a notation, somewhat unclear, to transcribe mathematical propositions; (ii) he did not develop a true *calculus raciocinator*; (iii) his viewpoint about axiom systems was antiquated from the beginning; (iv) he plagiarized his celebrated five axioms from Dedekind and his idea of quantification from Frege; (v) he ignored relations in his logic, in which he had to be 'corrected' by Russell; (vi) he lacked a philosophy of mathematics; (vii) he did not know the analysis in terms of truth values; (viii) he rejected Frege's (and Cantor's) logicism. As we shall see, all of these statements are very open to doubt.

I shall begin with a study of Peano's logic (3.1) and, once the central ideas of his arithmetic and his geometry have been pointed out from the logicist viewpoint (3.2, 3.3), I shall end with a global characterization of his method (3.4). Finally, I shall include a brief study of the various improvements proceeding from the contributions of his followers to Peano's general system, which were decisive for the main topic of Russell's method: the role of nominal definitions (3.5).

---

[1] Peano's only achievements usually mentioned, apart from some articles, are (a little part of) his famous notation and his five axioms for arithmetic. I know only three wide studies on Peano: Jourdain *1910b* (the best one, it is precise and accurate, although I think too influenced by Russell's PM); Terracini *1955a* (a collection of short papers mainly in the well-known style of a *Festschrift*); Kennedy *1980a* (a summary of the former papers by this author in the framework of a biography; one problem with this book is that it contains very little references to the works by Peano, and this mainly from the mathematical viewpoint). There are some more relevant studies, but they are limited to the extension of an article (Vailati *1899a* and Couturat *1899a* are to be emphasized because they were contemporaries to Peano). The articles by Ugo Cassina (the celebrated editor of the *Opere Scelte*) are, of course, also interesting, but they seem to be too influenced by Peano himself and lacking a philosophical approach. Two recent useful articles are Grattan-Guinness *1986b* and Quine *1986a*. (Added in proofs: Now I receive the book M. Borga, P. Freguglia, D. Palladino, *I contributi fondazionali della scuola di Peano*, Milano: Angeli, 1985, which is a good, global introduction.)

3.1. Logic

3.1.1. Objective and stages

Peano described himself as 'incompetent' in philosophy (*1891e*, 115; Geymonat *1955a*), but he reached, as we shall verify, a clear and efficient vision of the great generality of his main project. A sign of this vision was the clarity with which he defined the goals of his logic: the formation of a symbolic notation and the study of the rules of transformation (or reasoning). Both things are deeply related: they proceed from a complete analysis of the ideas and reasonings actually used in mathematics in general, including logic. This analysis led him to the fundamental logical ideas and identities that allow to define and demonstrate, respectively, the rest of concepts and propositions (*1894c*, 257; *1894e*, 286). Therefore Frege's claim according to which Peano's logic was a *lingua characteristica*, rather than a *calculus raciocinator* (quoted in Largeault *1970a*, 29-30), has to be regarded as out of focus. Quine regards (personal communication) the accusation as 'reasonable' but, as we shall see below, for Peano both aspects are not (and they cannot be) independent.

It is undeniable, however, that the immediate usefulness of the logical symbolism, as an instrument to express mathematical propositions, dominated Peano's efforts. In fact, we can find here the seed of Russell's (and partially Frege's) 'empirical' attitudes towards logic, characterized by considering it as a justification of mathematical expressions rather than as an end in itself (together with a lack of a clear distinction between primitive propositions and rules of inference). Thus logic would be useful mainly to *express* and demonstrate in a rigorous and concise way a set of propositions and theorems which would take much more space in natural language (*1890c*, 141): 'the usefulness of symbols comes to light and is measured by its applications' (*1900a*, 310). Besides, Peano was convinced, against Frege's referred accusation, of having given a 'single' solution to Leibniz's twofold problem, for according to him mathematical logic refers to (and develops) a set of *truths* in themselves and not a mere set of conventions (*1896a*, 197), with which his view was very similar to that of Frege himself. Consequently Peano soon came to occupy himself as well with logic as a science deserving of its own study.

Already in 1889, after having offered both foundations of arithmetic and geometry, Peano realizes the need for making the involved logic explicit, by overcoming Boole's attempts and the work already made in *1889a*; thus, he writes that it would be interesting 'to distinguishing the fundamental [ideas], which have to be immediately admitted, from the rest contained in them', which would constitute a study similar to the one already carried out for geometry and arithmetic (*1889b*, 81). This first attempt led him in 1891 to his first systematization of logic and to the first *explicit* distinction between primitive and derivative ideas into a science (*1891c*, 204). As we shall see, this was the beginning of an entire evolution whose final stage was not very satisfying because of the difficulties which arose in the development of the plan. At the end, Peano even admitted that logic (at least that of the *Formulaire*) must not be seen as a science in itself (*1912b*, 284). In the meantime, he set the ground of the contemporary mathematical logic.

This entire evolution can be summed up in five stages (Cassina *1933a* points out only three). In the first one (*1888a*) Peano follows Schröder's calculus of classes and he points out the duality with that of propositions. In the second (*1889a*), the new 'ideography' appears, which is applied to arithmetic taking the calculus of propositions as a basis. The third (*1892c*) already offers a study of logic in itself (also starting from the calculus of propositions), and it includes *1894b* and *1895a*. The fourth stage (*1897a*) provides the most extensive and ambitious theory, now starting from the calculus of classes (which will already be definitive), and attempts the reduction of primitive ideas and propositions to the smallest number (which leads Peano to discover the interdefinability of the first ones). Finally, in the fifth stage (*1900a*) the reductive effort is minimized and a trend begins to transform logic into an auxiliary instrument rather than an object of investigation in itself, until arriving at the last version of the *Formulario* (1908), where logic already occupies very few pages. The continuous oscillations between classes and propositions seem to be the result of basic philosophical problems; however, in 1912 (several years after abandoning logic) Peano attributes them to the successive pre-eminence of two different criteria: the capacity to be extended (propositions) and the rigour in the exposition (classes), although with little conviction: 'The order of propositions is uninteresting' (quoted in Jourdain *1910b*, II, 273).

3.1.2. Primitives, logical order and interdefinability

Already in *1889a* Peano was aware of having introduced a genuine method, based on the distinction between *primitive* ideas and propositions (*Pi, Pp*) and those being *derivative*. However he then did not employ such terms, speaking about signs representing undefined ideas (which can be used to define the rest of them) and undeduced propositions (postulates or axioms), which express the properties of undefined ideas (*1889a*, iv). The application of this to geometry provides the same distinction, but the status of the primitives is somewhat more clarified in establishing the primacy of the formal relations between them (*1889b*, 77), i.e. in pointing out the notion of interpretation as a mechanism capable of providing meanings that can (or cannot) satisfy the axioms.

In *1890c* (p. 128) the idea of function is claimed as 'primitive': 'It can be regarded as belonging to logic'; but in spite of the 'logicist' viewpoint identifying what is primitive with what is purely logical, it seems to be an unrepeated lapse. Only in the next year the explicit distinction may be found: 'The ideas that appear in a science are distinguished into *primitive* and *derivative*, depending on whether they can or they cannot be defined' (and the same for propositions) (*1891c*, 102). Already since then the relativity of the primitive character of a particular idea is clarified, together with the distinction itself: 'The distinction of the ideas of a science into primitive and derivative is sometimes arbitrary', therefore their use rests only on 'reasons of simplicity' (*ibidem*; this idea appears already in *1889b*, 78).

The application of the method already used in geometry and arithmetic to logic is emphatically valued: for Peano it is 'the first one of its kind' (*1891c*, 105). Today we know that Frege and

Peirce anticipated him in this area; in fact a certain indirect influence from them through Schröder cannot be excluded, for Peano himself recognizes that in Schröder's works there are also primitive ideas and propositions, although he also emphasizes the different viewpoints with regard to the ordering, the axioms and the definitions (*1891e*, 115). In any case the basic distinctions, if not the terminology, were already present in 1889, i.e. before the publication of the first volume of Schröder's work in 1890. Therefore it is necessary to regard rather Grassmann and Pasch as important influences, mainly due to their achievements in the constructions of formal systems explicitly containing axioms (see Freudenthal *1962a*, Kennedy *1973a* and Botazzini *1985a*). However, it is undeniable that Peano knew since *1889a* some of Peirce's works, for he mentions him in the bibliography (in 3.1.5 we shall return to Peirce).

According to Peano, the practical way to obtain the primitive ideas and propositions of a science is to replace the logical terms for their symbols and to analyze the rest of terms until finding the simplest ones (*1894b*, 164). This will provide those ideas with which all the others can be defined (though, as we shall see, such a 'simplicity' can be arbitrary). In any case, simple ideas are the common heritage of all human beings, and they are acquired by experience, not by deduction or definition; exactly in the same way as the rules of logic. The relation between Pi and Pp is that the second ones offer the fundamental properties of the first (*1894b*, 173-5). At this point we can already see the great coincidence with Russell's ideas before the Paris Congress.

The subsequent process of reducing them to the smallest number should consist in going deeply into the one already employed for obtaining them; we have to look for definitions (and demonstrations) with which we can come to express (or demonstrate) most of the ideas (or propositions) involved. In *1897a* six Pi (named as 'notations') are given: $\in$, $\supset$, $\cap$, $=Df.$, the letters a, b, etc., and finally the symbol k (class). To these ideas two (more problematic) others have to be added: the notion of 'pair' ((x, y)), which is introduced through a Pp stating the equality of pairs: $(x, y) = (a, b) \ . = . \ x = a \ . \ y = b$ (*1897b*, proposition 70); and the idea of negation, whose difficulties, on being introduced (like Peirce) through other Pi led Peano to determine its meaning by means of three Pp (*1897b*, propositions 105-7). However, only three years later he explicitly wrote that negation is another Pi, 'whose value remains determined by the primitive propositions...' (*1900a*, 341). On the other hand, the notion of pair is here added to the list of 'notations' or primitive ideas.

Peano insisted, even in *1898a*, on the reduction, eliminating two of the logical Pp and offering a foundation of geometry with only three Pi (*1898c*); but he also found many possibilities of interdefining the logical Pp, so that his first enthusiasm for the distinction (primitive-derivative) itself decreased. This made the difference between the *real* definitions (those reducing an idea to another within the chosen logical order) and the merely *possible* ones (those referred to other possible logical orders and to other Pi) vanish. That is why Russell wrote, in a somewhat exaggerated way, that Peano had renounced to the attempt to emphasize any kind of Pi from 1897 on (POM, §32). The truth is that a certain relativism undermined the distinction and led Peano to the pre-eminence of the 'implicit' character of the definitions of the Pi through a set of Pp ruling them (*1900a*, 319). In 3.2 we shall describe a parallel relativism in arithmetic.

Consequently, the evolution of the distinction primitive-derivative was linked to that of the notions of logical order and interdefinability, which can be understood if we remember that only the *actual* processes of defining and demonstrating could state a particular logical order. In 1889 we can speak of logical order if we refer to the fundamental works (on arithmetic and geometry), since in them there is a distinction between undefined ideas, definitions, axioms and theorems. However concerning logic itself, only the 'notations' are given, together with a long set of propositions (*1889a*) founding the calculus, but without an explicit clarification about which of them must be regarded as definitions, axioms or theorems (and even without the appearance of demonstrations, which are provided only for arithmetic). Perhaps Peano thought that logical propositions should be obvious, in being an absolute starting-point. In any case we already find some instances of interdefinability:[1]

$$a = b .=: a \supset b . b \supset a; \qquad a \cup b .=∴ -:- a - b; \qquad a \supset b .=. a - b = \wedge;$$
$$a \supset b := : x \in a . \supset_x . x \in b; \qquad a = b := : x \in a .=_x . x \in b.$$

That is why, as soon as Peano had the first distinction between primitive and derivative ideas in logic available, he was forced to recognize that the reduction of a sign to others causes two problems: (i) the arbitrariness of any set of selected ideas; (ii) the need for finding a criterion of simplicity (which, again, points out the problematic independence from ordinary language, mainly in such drastic cases like that of identity as mutual implication; *1894b*, 136). Below I shall return to these problems.

It is possible that the correspondence with Frege (and their unfortunate dispute on which of the two systems exhibited the smallest number of primitive symbols)[2] led Peano to attempt a reduction to the smallest possible number of *Pi* in *1897a*, especially if we notice that in his response to the famous letter from Frege he admits that the number of his primitive symbols is greater than he thought, and he adds that his reduction is not the 'ultimate word' (*1898b*, 295 f).

We find some already very elaborated examples of reduction in *1897a*, chiefly through the class abstractor, the new symbol $\bar{\iota}$ (later $\gimel$) and quantification (pp. 212, 216):

$$\sim a = \overline{x \in} (b \in K . a \cup b = \vee . \supset_b . x \in b) \quad Df;$$
$$\wedge = \bar{\iota} K \cap \overline{a \in} (b \in K . \supset_b . a \supset b).$$

---

[1] They are examples of the calculus of propositions (propositions 3, 23, 39) and of the calculus of classes (propositions 50, 51), all from *1889a*. The first one defines the equivalence in terms of implication, but it did not appear there for the first time (as affirmed by Jourdain *1910a*), but in *1888a* (though in terms of inclusion).
[2] Letters of Frege and Peano (1896), in Frege *1976a*, 113 ff and 118 ff.

This leads to the relativization of the chosen set of *Pi*: 'We can try to reduce further the number of the ideas regarded as primitive, or to try another way, assuming another set of ideas as primitive ideas so that another type of simplicity is obtained' (*1897a*, 217).

The consequence is the appearance (in *1897b*) of the symbol [*Df*], which is added to those propositions that could be taken as definitions 'by changing the system of primitive ideas' (*1897b*, 271) (this symbol was later replaced by *Df?* in 1900, and by *Dfp* —possible definition— in 1901). This allowed the distinction between genuine definitions, or reductions (*1897b*, propositions 201, 400), and the 'mere [*Df*]' (*ibid.*, propositions 52, 241):

$$a \cup b = \text{-}[(\text{-}a)(\text{-}b)] \; Df; \qquad \exists a \,.\, = \,.\, a \text{-=} \wedge \; Df; \qquad a \supset b \,.\, = \,.\, a = ab \; [Df];$$

$$a \cup b = \overline{x \in (c \in K \,.\, a \supset c \,.\, b \supset c \,.\, \supset_c .\, x \in c)} \; [Df].$$

However the power of such a recourse had devastating consequences for the chosen logical order; not only because of the obvious fact that many possible definitions allow defining, e.g. certain ideas in terms of the rest of the ones chosen as primitive through certain recourses, but because in some cases the *conceptual* foundation of the system would be affected, along with the meaning of the symbols. Thus for example, Peano emphasized that the idea of deduction ($p \supset_x q$) can be reduced to logical addition and negation (like McColl had already seen: $\text{-}p \vee q$), although then, when we do not write '$\text{-}p \cup q =_x \vee$', one of Peano's ways of expressing quantification vanishes (*1897b*, 266). We have another example in defining inclusion through existential quantification (*1897b*, proposition 413):

$$a \supset b \,.\, = \,.\, \text{-}\exists (a \text{-} b) \; [Df]$$

where a relation ($\supset$) comes to be expressed as a property ($\exists$). Something similar occurs when Peano realizes that the concept of class can even be reduced to that of property (*ibid.*, 276), and when he verifies that on reducing negation to symbols previously defined in terms of the *Pi* (*1897b*, proposition 433):

$$\text{-} a = \iota K \, \overline{x \in [a \cap x = \wedge \,.\, a \cup x = \vee]} \; [Df],$$

it is shown that 'the idea of negation is not primitive'. This would force us, if we eliminate it, to a change in the *order* of definitions, of *Pp* and of all demonstrations; i.e. to create a new calculus (*1897b*, 270).

Therefore Peano openly recognizes, not only that the chosen order is relative, but also that both for the *Pi* and the *Pp*, he has chosen the ordering that 'seems to him' the simplest one, by indicating other possibilities through the symbol [*Df*] (*1897b*, 247). Consequently, the power of eliminating definitions is already not limited to the defined symbols, but also to some of the chosen undefined ones: 'Thus, for example, the four symbols ⊃, =, ∧, ∃, are reduced to one another so that every form of reasoning is found written four times in the *F[ormulaire]* under different forms' (*1898a*, 287). However, the only offered reasons for maintaining all of them is the mere convenience of arranging them in certain contexts; although with no kind of defence of any privileged logical order, which is recognized as being possible in many ways (*1898f*, 301). As we shall see, one of the basic consequences of this view is that of suggesting the possibility of altering the established order into other chains of definitions, as for example into those of mathematics, with which the borderlines between the traditional subjects would perhaps be modified. Peano admits this consequence (which will be vital for his implicit 'logicism' and his influence on Russell) affirming that arithmetic, algebra and theory of numbers are already not sciences with definite boundaries: 'That which is called "logical order" is often nothing but a more or less established habit' (*1898d*, 241).

From 1900 on all the technical consequences are already drawn (*1900a*, 310): the usual definitions and, therefore, the *Pi*, are always relative to a given order; we must choose, as real definitions, the most handy ones, and that logical order where the smallest number of *Pi* takes place. From the philosophical viewpoint this leads him to the curious conclusion that what is absolute or intrinsic rests no longer on the *Pi* (and the *Pp*), since the 'real' definitions are now always relative, but on the 'possible' definitions (now through the symbol *Df?*): 'The symbol *Df?* expresses, then, an intrinsic property of a proposition; the symbol *Df* a property which is relative to its place in a theory' (*ibidem*; see however 3.2 below). That is why starting from this point Peano more emphatically claims that the set of *Pi* can *be defined* as the system satisfying the *Pp*. Thus, the true logical order is a structural problem rather than a linear sequence of definitions.

3.1.3. Implication, inclusion and membership

Peano introduced the symbol ⊃ for inclusion and implication and explained that the two possible readings would depend on the respective uses between classes or between propositions. The possible ambiguity is solved through the remark (perhaps inspired by Frege) that the two interpretations do not prove that the symbol has several meanings, but only that there are several terms in the ordinary language that represent the same idea (*1897a*, 208; letter to Frege from October 1896, in Frege *1976a*, 121). Before the possibilities of reducing the symbol to an identity (for instance $a \supset b . = . a - b = \wedge$) or to logical addition and negation ($a \supset b . = . - a \cup b$), Peano chooses the viewpoint of ordinary language, arguing that the acceptance of these possibilities would lead us too far from the common usage, and also from the practical advantages of having a specific symbol available.

Peano clearly said that the best way to understand the meaning of a symbol is the analysis in terms of truth values, but he also pointed out that this form of analysis is only valid when the symbol appears between propositions that do not contain variables (or undetermined letters) (*1894b*, 139). This seems to introduce the distinction between material and formal implication (the terms used by Russell in POM): the first one would take place between propositions with a constant value and the second between propositions with variables ('propositional functions' according to Russell). The problem is that, as Peano expressed universal quantification through a subscript (to the symbols for implication and identity), he tended to give the pre-eminence to the 'formal' type of implication. Hence he criticized Frege for expressions like $2 > 3 . \supset . 7^2 = 0$, where the implication appears between constants (*1897a*, 207), despite that he himself had used examples like $1 > 2 . \supset . 5 > 4$ (*1894b*, 139). Therefore, when Peano writes: 'the symbol $\supset$ is essentially used by us between propositions containing variable letters' (*1897a*, 207), he is not excluding the other use, but stating that only this 'formal' implication reaches the greatest generality, in being especially constructed to be valid in *all* cases, independently of the value of the variable (whenever its field of values is previously specified). Thus, as the truth of a variable proposition depends on the value that it takes, Peano preferred to restrict the analysis in terms of truth values to the case of the implication between constants. In any case the distinction seems clear, like the implicit one between proposition and propositional functions (this against Jourdain *1910a*, II, 219).

As for the distinction between membership and inclusion and the invention of a specific symbol for the first one, it is one of Peano's greatest achievements, which was introduced before Frege did it.[1] As we shall see in 3.1.5, to express mathematical propositions requires a combination of membership and implication allowing the apprehension of the generality of the variable. The other basic capacity of membership is the conversion of the calculus of classes into the calculus of propositions. Curiously, when it was introduced for the first time in *1889a*, some examples show a certain bad use as regards inclusion, for at times the symbols of membership appears between classes[2] (in cases where classes are regarded as individuals, starting from the more confusing reading 'is').

A few years later, however, a whole set of properties appears, which set the distinction membership-inclusion more carefully (*1894b*, §16). Finally Peano emphasizes four basic differences (in the review of an article by Schröder, who did not distinguish them): (i) transitivity ($\in$ does not have it); (ii) commutativity (of $\in$ and -); (iii) distributivity (of $x \in$ and $\cup$); (iv) existential presuppositions (if $x \in a$ then $\exists a$; which is not true for $a \supset b$) (*1898f*, 299-301). Although, just like the symbol of inclusion/implication was shown dispensable on being defined through others, Peano soon discovered the way to eliminate membership, as a supposed primitive idea, in terms of the unit class and the product of classes (*1897b*, propositions 422, 424):

$$x \in a . = . \iota x \supset a; \qquad x \in a . = . \exists (\iota x) \cap a.$$

---

[1] It did not appear in Frege *1879a*, as Russell wrote in POM; see Jourdain *1910a*, II, 269, and Kennedy *1980a*, 26.

[2] Like propositions 52-56 of *1889a*, and mainly axiom 9, where we find $k \in K$.

But also here the practical value of the symbol, resting on its role in ordinary language, permitted its preservation.

3.1.4. Classes, propositions and individuals

In the five stages of the evolution of Peano's logic mentioned above, I pointed out the successive oscillations of the respective pre-eminences of classes and propositions. I also mentioned the role of $\in$ in the conversion of categorical propositions into conditional ones. However it was the introduction of the class abstractor that made the application of the symbols of logic to propositions possible. Already in *1888a* it was introduced as $x : \alpha$ (the class of all entities for which the proposition α is true), although this symbol was replaced by $[\in]$ in 1889 and later by $\bar{\in}$ (the converse of $\in$) and $\ni$. The important thing is how, starting from a proposition containing a variable, the class of all x satisfying it may be defined (*1889a*, xii, proposition 57):

$$a \in P . \supset : [x \in] a . \in K.$$

Thus, a property of $\in$ (to transform inclusion —a relation between classes— into implication between propositions) and a property of $\bar{\in}$ (to transform implication between propositions into inclusion between classes) are joined (*1889a*, propositions 50, 63):

$$a \supset b . = : x \in a . \supset_x . x \in b; \qquad \alpha \supset_x \beta . = . [x \in] \alpha \supset [x \in] \beta,$$

which permits him to express, for example, operations between classes in terms of operations between propositions (*1891c*, §5, proposition 6):

$$a \cap b = \overline{x \in} (x \in a . x \in b)$$

In 1894 the parallelism was emphasized, especially through a reciprocal foundation of both calculi. Here are some examples of the correspondence (*1894b*, §§16-7; I dispense with the subscript 'x' to propositions —to indicate they contain variables— in order to simplify):

$$x \in a \cap b . = : x \in a . x \in b \qquad \overline{x \in} (p \cap q) = \overline{x \in} p \cap \overline{x \in} q$$

$$x \in -a . = . x - \in a \qquad \overline{x \in} (-p) = -\overline{x \in} p$$

$$a = \wedge . = : x \in a . =_x \wedge \qquad p =_x \wedge . = . \overline{x \in} p = \wedge.$$

However, starting from the pre-eminence of classes in 1897, Peano only pointed out both possibilities in terms of reduction (*1897b*, comments to propositions 12 and 70).

Russell (POM, §13) criticized the duality (following Couturat *1899a*) by pointing out particular instances where it was not fulfilled; in particular in searching for examples where implication cannot be transformed into inclusion (also following Schröder). With that he intended to justify his choice of propositions as a more logicist ground for logic, following McColl and Peirce (although he described as 'obscure' the ultimate relation between both calculi). However Peano himself was aware that, no matter the practical advantages of each calculus, the limitation of the duality of the symbol lies in that, if we take it as a *Pi* for a calculus of classes, all the hypotheses preceding the propositions (in order to state the meaning of the letters) have to be linked to them through an *implication*, with which the idea would implicitly be no longer primitive. For example (*1900a*, §1, *2):

$$a, b \in Cls \;.\; \supset \;\therefore\; a = b \;.\; = : a \supset b \;.\; b \supset a.$$

Hypothesis

Besides, as Vailati already pointed out much earlier (*1899a*, 96), though the calculus of classes subsists if it rests on that of propositions, we do not have the same situation according to the converse possibility: the relation $a \supset b$ between propositions is a proposition, whereas the same relation between classes is not itself a class, but another proposition. Peano probably was aware of the problems which arose by any attempt to define the notion of proposition or to introduce propositional functions (which Russell had to admit as hardly distinguishable from classes themselves): all these difficulties could be avoided through the rigour of classes, whose operations are similar to those of algebra.

The distinction membership-inclusion, in emphasizing the existing differences between the individual and the class, should have contributed to make the existing one between a class and the sum of its members easier (which, for the unit class, is equivalent to the distinction between an individual and the class of which it is the only member). However this last distinction appeared in Peano as an analysis of the symbol for equality (*1890c*, 130; on dealing with the inversion of functions):

$$(a = b) = (a \text{ is equal to } b) = (a \in \iota b).$$

Through the symbol ι it was immediately made possible to distinguish between a class in itself and the same class regarded as an individual. With such a device classes of classes could be managed without limitation (*KK*, *KKK*, etc.), but Peano must have already foreseen certain dangers in this process, so that he wrote that *KK*, *KKK*, and so on, can be regarded as individuals, 'but one must soon stop at ordinary language and at the applications' (*1890c*, 131). The practical usefulness is obvious as soon as we want to distinguish a class as individual (ιa) from a class in itself (a). Peano's example is illuminating: if a and b are straight lines, the expression $a \cup b$

designates the whole set of points, whereas $\iota a \cup \iota b$ designates the pair of straight lines as individuals.

However, the main role of ι was not the handling of the classes of classes (to which the logical sum and product were soon defined: *1894b*, §21; *1897b*, 272), but the already mentioned analysis of equality and the possibility of having a symbol available which would permit him to clarify the relation between class and individual. In Peano's own words: '[ι] is then a symbol of function that, written before any individual x, gives rise to a class, ιx, the class of the individuals that are equal to x' (*1894b*, 160). Starting from there, the distinction between the 'is' of membership and that of identity becomes clear, the confusion of the unit class and its only member is avoided, and the definitions of the notions of individual and class become unnecessary (*1894b*, 131): they will be, respectively, all that appearing to the left or to the right of ∈.

In 1897 the gradual process arrives at its end by defining ιx as the class formed by any *one* object and by introducing the converse $\bar{\iota} a$ as the *only individual* of the class (*1897a*, 215, definitions 21-22; *1897b*, 234, propositions 420, 430):

$$\overline{\iota x = y} \in (y = x); \qquad x = \bar{\iota} a \ . = . \ a = \iota x$$

'a' being an existent class and 'x' its only member. The new symbol $\bar{\iota}$ made the definition of many others possible, especially through its capacity to isolate an entity, which was very useful at that time, when Peano was trying to reduce the primitive entities to the smallest number. The only change would be to designate the same concept by another new symbol: γ (*1900a*, 352), which was later used by Russell with very important consequences for the theory of descriptions.[1]

3.1.5. Mathematical propositions and quantification

Peano's supreme ambition was always to achieve a system for *expressing* all propositions of mathematics. For this he needed a logical symbolism and an 'analysis' of the mathematical ideas, but mainly a mechanism that would allow him to express the great generality of mathematics and the implication between general propositions. The two first devices were achieved in 1889. The last one was already in McColl and Peirce, with their distinction between categorical and hypothetical (conditional) propositions and their pre-eminence of the calculus of propositions based on implications, which made it possible to overcome the equational logic of Boole and his followers (see 1.1 above). The essence of the distinction lies in something very simple: the primacy of the conditional (also pointed out by Bradley; see 1.4 above). Thus, a categorical judgment like 'all a are b', is transformed into 'if x is an a, x is a b'. In fact, the distinction was

---

[1] The conditions to make use of this symbol were *existence* and *uniqueness* (*1897b*, 268-9; also in Frege *1884a*), so that a great part of Russell's theory of descriptions seems to be underlying here (see my *1989b*, *1990h* and *1990i*).

already present in Peano *1888a* (p. 8); with it, the calculus of classes can rest on that of propositions through the symbol of membership (see 3.1.4 above). This latter sign had, besides, the capacity of transforming a common proposition into another containing undetermined entities, which is the seed of the distinction between propositions and propositional functions.

As for inclusion, its mere introduction turns the equation $AB = 0$ into $A < B$, which later came to be written as $A \supset B$, giving rise to the primacy of the conditional (the 'inclusion' between propositions) already since 1888. Concerning the class abstractor, we saw before that it is also already present in 1888 as the converse of $\in$, and it was interesting because it made the application of the logical symbols to propositions possible.

The only thing lacking is quantification, which soon led Peano to the distinction between real and apparent variables (or 'letters'), already implicitly present in 1894,[1] and fully in 1897: 'It is said that, in a formula, a variable letter is apparent if the value of the formula is independent from the variable letter' (*1897a*, 206; also in *1897b*, 243). (As it is well-known, the distinction was inherited by Russell and gave rise to the present one between free —real— and bound — apparent— variables.) The link with universal quantification was also clarified by Peano when he said that all variables in a proposition containing $\supset_x$ are apparent variables (just like any variable contained in a theorem). Below I shall return to quantification.

All of this set of devices, together with a method for definitions and the idea of an axiom system (see 3.4 below), necessarily had to impress Russell (as can be seen in his *1900a*, 6). The 'essence' of the variable, which he was looking for for a long time (in the manuscripts), appeared in all its generality and efficiency through the simple combination of $\in$ and $\supset$ (e.g. $x \in N . \supset . \ldots$), with which the word 'any' ceased to be a problem. Sometimes Peano has been criticized (Jourdain *1910a*, II, 299) with the argument that he did not explicitly mention that which allowed the expression of mathematical propositions is precisely the propositions with variables and the implication between them, and even that he confused propositions and propositional functions. As we saw in 3.2.4 this argument rests on the supposition that Russell *discovered* 'formal' implication, but this supposition does not seem compatible with the facts (apart of the mere terminology).

I now come to the historical priority of the quantification device. It is usually accepted that Frege and Peirce (independently) discovered it, and that Peano took it from Frege, who had the concept available from *1879a* (e.g. Bynum *1972a*, 41). It is true that we find the explicit recourse to quantification in *1894b* (§14, 138), through the adding of subscripts to the symbols of implication and identity (and the same in *1894a*, 120, 122). Likewise, we find the symbol $\exists$ for the first time in 1897, but only as a mere abbreviation of the old idea: $\exists a . = . a - = \wedge$ (*1897b*, proposition 400; also in *1897a*, definition 19), which of course is attributed to Frege *1879a* (*1897b*, 266).

---

[1] In particular in *1894b*, 137. There we read, after some examples in which the values of the variables do not depend on the letter, that 'dans ces formules il n'est plus nécessaire d'expliquer la signification de x, car cela est déjà dit dans la formule même'.

However, all of these ideas are much earlier. Already in *1891e* (a review of the first two volumes of Schröder's *magnum opus*) Peano writes that Schröder's notation for quantification (taken from Peirce), i.e. $\Sigma_x a$ and $\Pi_x a$, can be written through subscripts to some symbols, like for instance 'a - $=_x \wedge$' and '$x \in S . \supset_x . a ...$' (*1891e*, 119, definitions 4, 9). But Peano made use of these notations even before the publication of Schröder's second volume in 1891 (the relevant one for dealing with the calculus of propositions), as may be seen in his *1890c* (p. 121).

Nevertheless, the two basic ideas were already present in the two fundamental works of 1889. Here are several examples:[1]

1. $a \supset_{x, y} b$ (for all x and y, if a then b)
2. $a =_{x, y} b$ (it indicates an identity with regard to x and y)
3. 1 - $= \wedge$ (the class *point* is not the null class, i.e. there *are* points)
4. 'the proposition $a =_{x, y, ...} b$ means the same as $a \supset_{x, y, ...} b . b \supset_{x, y, ...} a$'
5. $a \in b . .... b$ - $= \wedge$ (it indicates existence, in spite of being referred to classes).

We must conclude, then, that Peano had available clear notations and ideas on quantification before knowing the relevant works by Schröder and also before knowing, as far as we know, Frege (he quotes Frege for the first time in *1891b*, 101). As in the case of the distinction between primitive and derivative ideas and propositions, the possibility subsists that he would take the idea of quantification from Peirce (whom he mentions already in *1889a*), who had made use of it from 1885 on (see Thibaud *1975a*, 96 ff), but, by the form in which he rushed to reduce Peirce's achievements into his own notation (in the mentioned review of Schröder) it seems quite unlikely. As for the correspondence with Frege (Frege *1976a*, 108 ff), it contains nothing that forces us to modify this conclusion (more details in my *1988b*).

3.1.6. Relations, functions, classes, properties and propositions

Peano's first works introduced separately the notions of class, function, relation and pair. The first one (as we saw above) would be turned into a basic one from 1897, in spite of the possibilities of an alternative calculus of propositions, or a reduction of classes to properties. As for the idea of function, it appears as a 'complement' to *1889a* and as a *Pi* from 1890 on, being mainly destined to give an account of the various types of correspondences and of the 'converse' notions ($\bar{\iota}$ or $\overline{x \in}$). The concept of *pair* (named later *ordered pair*) also appears in *1889a* (perhaps coming from Peirce, though Frege and Schröder made use of it too), being explained as 'a new entity' (*1889a*,

---

[1] The examples proceed from: (1) *1889b*, 59; (2) *ibidem*; (3) *ibid*., pp. 64, 83 (axiom 1); (4) *1889a*, viii; (5) *ibid*., xi (proposition 53).

xii). In 1897 the notion reappears, but now is 'defined' through the equality of pairs, emphasizing its primitive character and mentioning the order of the elements: 'the pair $(x\,;y)$ is also said to be equal to $(a\,;b)$ when their elements are equal and equally ordered' (*1897a*, 213-4).

Relations incidentally appear in 1889 through several examples (equality, greater, etc.). Peano looks for a notation expressing them through a class and a property (*1889a*, xii); thus, he introduces the symbol [∈] (or ∋), and, starting from a relation $x\alpha y$ between any entities, he designates through $\ni\alpha$ those x satisfying $x\alpha y$; which is equivalent to say that $\ni\alpha$ are those x having the property αy (e.g. to be lesser than y). It is then an attempt to dispense with relations or, at least, to decrease their ontological import.

This same reductive line was strengthened by resorting to the concept of function: if we start from a relation $x\alpha y$, then, given y, a class of x satisfying this relation is determined; that is why 'this class is a function of y'. If we designate it by φy, we would have $x\in (x\alpha y) = \varphi y$; consequently $x\alpha y = x \in \varphi y$, with which a relation has been *analyzed* (reduced) into the symbols of membership and function: $\alpha = \in \varphi$ (*1894b*, 159-60) (the same device is used to analyze equality; see 3.1.4 above). The definitive reduction was, however, that of relations into classes of pairs (*1897b*, 256): 'every relation between two objects x and y is reducible to the form $(x, y) \in a$, "a" being a particular K [class] of pairs'. From here on the idea continued to be developed.

We now come to the relationship between classes and functions. Peano had made efforts to distinguish classes and operations (functions, correspondences), on realizing that the latter ones can give rise to the former (just like properties or propositions can give rise to classes), in the same way that a function of two variables can give rise to a class of pairs, i.e. a relation. (This is very interesting if we remember Russell's efforts in POM to distinguish between classes, propositional functions and relations.) The most serious attempt proceeds from 1894 (i.e. before assimilating relations to classes of pairs in 1897): 'It is necessary to distinguish the names of classes from the names of operations', as it can be seen in propositions with no meaning such as 'x is a multiple', where 'multiple' is the name of a function and not of a class (*1894b*, 146). The problem is worsened with functions of two (or more) variables, which can be interpreted as such functions, but also as relations and as classes of pairs.

In considering such a situation, we can easily imagine Peano's surprise on receiving from Russell his first work on relations (*1901a*), where these entities were made explicit. In his reply (March 1901; in Kennedy *1975a*) Peano was forced to remind Russell that relations 'are' classes of pairs, pointing out his *1901a*. In 1904, nevertheless, he was more tolerant with POM, where Russell apparently regarded functions as relations. That is why he wrote (Couturat *1904c*, 1046) that it was the same to regard them in this way (as in his *1901a*) than to define relations in terms of functions. (The truth is that Russell regarded functions as primitive in POM; see 4.2 below.)

The immutability of Peano's view is reflected in his reaction to PM (*1911a*, 364; *1913b*, 394), when he, not only said once more that relations can be defined as classes of pairs, but showed particular definitions where functions are reduced to relations of a special type. However, Peano had no special interest in maintaining the notion of class as a *Pi*, as it is shown when he himself sometimes offers the possibility of reducing it, for example, to that of property (*1897b*, 276).

We must therefore point out the lack of philosophical reasons for Peano to choose this or that notion as primitive outside any practical purpose. Furthermore, it can be said that the five notions we have examined here were, for him, almost reducible among themselves. However, Russell needed, to maintain his general ontology, to give the pre-eminence to relations. It does not seem strange that he criticized Peano with the same argument he gave against Bradley: to have held the subject-predicate pattern (now on the only ground of Peano's celebrated thesis according to which all propositions *can be* reduced to the form $x \in a$; Peano *1900a*, 337; also in *1901a*). What Peano probably meant was that propositions are reducible to classes, in the same way that *relations* are reducible to the form $(x, y) \in a$ ('*a*' being a class). But, as we have seen, for him there were many other possibilities of inter-reduction according to our particular and practical purpose, so that our philosophical reasons cannot show any of these notions as *the* truly primitive.

3.2. Arithmetic

3.2.1. The axioms and their interpretation: Dedekind

Peano's arithmetic is the setting-up of some *Pi*, some *Pp*, and some definitions about integers, from which the usual operations are subsequently deduced, as well as rationals, reals, analytic functions and complex numbers. The first attempt took place in *1889a* (§1), where Peano set up, making use of a symbolic language, four *Pi* (number (N), one (1), successor ($a + 1$) and equality (=)) and nine axioms giving their relations. Two years later, Peano carried out an attempt (*1891d*) to simplify the axioms by introducing the concept of correspondence (mapping) and by eliminating four of them as belonging to logic (only the second change was maintained).

In *1898d* Peano starts from three *Pi*: 0 (instead of 1), number (N_0) and 'successor of' ($a +$), and the following five axioms:

1. $0 \in N_0$
2. $a \in N_0 . \supset . a+ \in N_0$
3. $a, b \in N_0 . a+ = b+ . \supset . a = b$
4. $a \in N_0 . \supset . a+ -= 0$
5. $s \in Cls . 0 \in s : x \in s . \supset_x . x+ \in s : \supset . N_0 \supset s,$

which are usually read: (1) zero is a number; (2) the successor of any number is a number; (3) no two numbers have the same successor; (4) zero is not the successor of any number; (5) any property that belongs to zero, and also to the successor of every number having the property, belongs to all numbers. The system suffered only minor changes.[1]

The five axioms can be found in Dedekind *1888a* in a somewhat different way (see 1.2 above), which was recognized by Peano (*1891d*, 86), though in 1889 he only mentioned the general usefulness of Dedekind's work in the preface (*1889a*, iv). However, though in *1898d* he transcribes the relevant passages by Dedekind immediately after offering the axioms (*1898d*, 218-9), Peano pointed out the independence of his discovery in a work of the same year: 'the composition of my work of 1889 was independent from the cited essay by Dedekind' (*1898e*, 243). He also adds that he knew Dedekind before submitting his own work to the printer, with which he felt more sure about the logical independence of the axioms, though he immediately tried to *demonstrate* that independence. There is no reason for rejecting this explanation, though it must be said that Pasch was probably the source of the idea of building an axiom system (see Kennedy *1973a*; *1980a*, 26, 174; and Freudenthal *1962a*, 617).

Peano's interpretation of his five axioms also was similar to Dedekind's. For Peano they express the necessary and sufficient conditions in order that 'the entities of a system can be put into one-one correspondence with the series of numbers' (*1891d*, 87), so that it is already implicit that many different systems can be 'interpretations' of them. That is why he adds an abstract characterization of the axioms, in a similar way to the usual definitions of abstract structures (already in his work on geometry of 1888 he had defined one of these structures: that of vectorial space).

Peano soon explicitly admitted the existence of many models: 'there is an infinity of systems satisfying all Pp' (*1898d*, 218), since all are verified if, for example, N_0 and 0 are replaced by N_1 and 1. As we saw above, for Peano this is possible for all systems that can be put into one-one correspondence with numbers. With that, Peano anticipated Russell's criticism (in POM) according to which Peano's axiom system for arithmetic failed to make a difference between the progression of numbers and other progressions. That is why Peano wrote: 'Number is what is obtained by abstraction from all those systems', i.e. numbers are the system having all and only these properties (*1898d*, 218; *1901a*, 44). As we shall see in 3.2.2, this statement was intended to be a certain kind of definition.

Sometimes it has been attempted to establish differences between the interpretations of the axioms made by Dedekind and Peano; for example (Aimonetto *1969a*), by pointing out that Dedekind began with any objects and emphasizing the interpretative and ordinal sides, whereas

---

[1] Three years later Peano added a new axiom: $N_0 \in Cls$, which was already implicit, since the symbol $\in$ requires classes on its right side (*1901a*, 5). As for 0, it was replaced by 1 for convenience (*1901a*, 39) and to avoid the difficulty to define 0 (*1901b*, 361). Curiously, Padoa had already defined such an annoying number: $0 = N_0 - N_1$ (starting from $N_1 = N_0 + 1$), which permitted the reduction of the number of Pp to only two (since equality has to be regarded as belonging to logic) (*1901a*, 44; see also the note by Peano in Jourdain *1910b*, II, 273). However, the subsequent editions of the *Formulaire* maintained the same axioms (including the last edition of 1908), in spite of the fact that they were demonstrated as theorems in the systems by Frege and Russell. This might be a sign that for Peano arithmetic must start with intuitive, indefinable notions.

Sometimes it has been attempted to establish differences between the interpretations of the axioms made by Dedekind and Peano; for example (Aimonetto *1969a*), by pointing out that Dedekind began with any objects and emphasizing the interpretative and ordinal sides, whereas Peano directly began with numbers and emphasized the cardinal, axiomatic and formal sides. However, as we have just seen, Peano's interpretation is inseparable from the intuitive 'abstraction' through which we come to numbers from a previous schema. Besides, though it is true that the *Pp* are formulable in terms of correspondence (like Cantor) and, therefore, can avoid ordinality, however Peano himself always preferred the traditional ordinal way of presenting the axioms.

Largeault (*1970a*, 428-9) has pointed out that Dedekind was convinced (in a clearer and more rigorous way than Peano) that his formal system was categorical (to use the modern term), i.e. that all of its models were isomorphic among each other. (However, although there are propositions which yield categoricity in Dedekind's writings, he did not seem to grasp this concept, as he allows for 'non-standard models' in his *1890a*.) But as we saw it seems that Peano was aware of this fact. However I should say 'fact', since after Gödel and Skolem we know that the internal problems regarding the notion of set make it impossible to obtain a unique characterization of the numerical series: any set of axioms for arithmetic is always polymorphic (non-categorical), i.e. the notion of set, implicit in the axioms, is shown to be as problematic as that of number.

I think Gillies' position (*1982a*, 66-8) is the true one. According to him the basic difference between them is that Dedekind tried to define the notions of arithmetic in logical (and informal) terms, whereas Peano intended 'only' to offer an axiom system to characterize them in a formal way. It is true that Dedekind can be regarded to some extent as a logicist for he thought that for introducing rigour into arithmetic this science has to be constructed as a part of 'logic'. However, we have seen above (and we shall return to it in 3.4) that, for Peano, only intuition granted the access to the basic concepts and axioms of arithmetic; that is why his axioms cannot be interpreted as an axiom system in the present-day sense. It is true that in a certain very precise sense (which I shall explain in 3.2.4) we can speak of 'logicism' in Peano, it is also true that, like Dedekind, he was convinced that his axioms *analyze* the notion of number, since they expressed 'the simplest properties of integers', whose concept resulted, thus, 'analyzed' (*1891d*, 97).

3.2.2. The definability of number

As we have seen, Peano thought that his attempt was parallel to that of Dedekind in the sense that, though the latter regarded his axioms as the explicit logical 'definition' of number and for himself they constituted 'only' its properties, they were dealing with the same thing: 'Here number is not defined, but the fundamental properties are enunciated. On the other hand, Dedekind defines number as that which satisfies the mentioned conditions. Obviously the two things coincide' (*1891d*, 88). However, Peano himself many times regarded such properties as a 'definition'. We integers, although of course adding that in this case the notion of definition is widened (*1898e*, 243).

In spite of this, it is undeniable that Peano insisted many times that the concept of number cannot be defined in logical terms. At the same time, he introduced the idea (of a great influence on Russell) that there is no sense in discussing whether a term (or an idea) is or is not definable without clearly specifying the available terms for the *definiens*. Already in *1891d* (p. 85) he set his definitive position in saying that *in practice* it is not suitable to define number: it is acceded to in a natural way, and no definition would improve our intuitive comprehension (e.g. in children). As for the *theoretical* possibility of doing so, he demanded that, in any case, the ideas to be used have to be clearly pointed out; and, if these ideas have to be those of logic, he added that then 'number cannot be defined' (i.e. an equivalent expression cannot be formed with those ideas). The five *Pp* just try to give the basic properties of this concept starting from a set of *Pi* which can be only obtained by 'induction'; and, if one wonders about the criterion of selection of the properties, only *simplicity* can be offered (*1891d*, 98), but this criterion is not applicable to the *Pi*, whose simplicity is superior to that of the ideas we could use to define them.

It is inferred from it that although *all* ideas of arithmetic, algebra and analysis were definable through logical ideas, it would be indispensable to add to these at least the three *Pi* from which arithmetic starts. In this direction obviously there are no traces of logicism in Peano; it seems even that he described Russell's famous definition as 'artificial' (Kennedy *1963a*; Geymonat *1955a*, 60). But Peano was completely aware (and this much earlier to knowing Russell's ideas) that *there is* another way to construct mathematics: that based on the notion of correspondence and on the definition 'by abstraction', i.e. Cantor's way. This way can also be regarded as logicist, as it is shown by the fact that Peano regarded it as equivalent to that of Frege, that is to say, as a possibility of 'defining' the cardinal number (*1895c*, 194 f). Before explaining this 'logicist' arithmetic, it is necessary to consider his construction of the reals, which anticipated some interesting traits.

3.2.3. Real numbers: construction and definition

Already in *1889a* (§9, p. 16) Peano included a theory of reals that, following Dedekind and Cantor, 'constructs' them as classes of rationals, though by means of the concept of 'upper term (or limit)' of a class, which was previously defined. Thus, a real number will be the (existing) upper limit of a (not empty) class (*Ta*) of rationals such that there are rationals greater than it (excluding 0 and ∞):

$$Q = [x \in] (a \in KR : a \text{ -} = \wedge : R \ni > Ta . \text{ -} = \wedge : Ta = x \therefore \text{ -} = \wedge).$$

But we find neither a 'philosophical' study nor a comparison with other constructions until *1891d* (§11), where he presents his own theory as an improvement of those by Dedekind and Pasch.

Dedekind's method is reduced to the identification of reals with the cuts produced by the two classes of rationals A_1 and A_2; and, as to define every cut it is sufficient with only one of these classes[1] (the first one being the 'rest' of the rationals from the second: $A_2 = - A_1$), we can identify reals through them. This device led Pasch to the notion of segment (*Strecke*), which Peano defined as an existing class of rationals not containing all of them, and such that if it contains one, it also contains all numbers lesser than it, and with no maximum. There remains only to identify reals with those segments.

For Peano, on the other hand, this identification cannot be carried out between the segment A and the number a. These entities can only be related by saying that a is the limit of A ($l\,'A = a$). Thus, reals will be the upper limits of the segments of rationals, and, once the concept of upper limit is defined, we shall be able to define reals, in general, as the upper limits of classes of rationals, excluding 0 and ∞ (*1891d*, 108):

$$Q = (l\,'\,KR)\,(-\iota\,\infty)\,(-\iota\,0).$$

Finally, though Peano points out the possibility of introducing successions (he mentions the construction by Cantor), this possibility is left aside as 'less simple than the previous ones'. In any case, the problem of the 'existence' is not posed, and it remains reduced to an obscure matter of simplicity (*1891d*, 109).

Since 1897 and the attempt to reduce the concepts to the smallest number, the definitive version appears: *1899a*, which offers, essentially, the same theory from 1889, but including notational improvements and an ontological discussion. In the new definition the real number is the upper limit of a non empty (existing) class of rationals (excluding 0) such that there are numbers greater than any number of the class (excluding ∞)(*1899a*, 253):

$$Q = x \in \{\,\exists Cls\,'\,R \cap a \ni [\exists a\,.\,\exists R \cap y \ni (y > l\,'a)\,.\,x = l\,'a]\,\}.$$

Concerning the construction of irrationals as segments (*1899a*, §§4, 8-10), it starts from the concept of proper fraction (θ) (*1899a*, 255):

$$\theta = R \cap x \ni (x < 1),$$

and from its application to rationals and the classes of them (*1899a*, 256):

$$a \in R\,.\,\supset\,:\,\theta a\,.=.\,R \cap x \ni (x < a);$$
$$u \in Cls'R\,.\,\supset\,:\,\theta u\,.=.\,R \cap x \ni [\exists u \cap y \ni (y > x)].$$

---

[1] This remark is usually attributed to Russell as a 'simplification' of Dedekind, but it can be found not only in Pasch or Peano, but also in Dedekind himself (*1872a*, 15).

Segments of rationals are then defined (*Sgm*) as the non empty (existing) classes of rationals not containing all of them, such that they contain all rationals lesser than any one of its elements, and such that each one of its elements is lesser than another of them (i.e. that they have no maximum) (*1899a*, 258):

$$Sgm = Cls'R \cap a \ni (\theta a = a . \exists a . \exists R - a).$$

Starting from this the former definition of reals is simplified; they are now upper limits of segments of rationals (*1899a*, 260):

$$Q = l\text{'}\text{'}Sgm.$$

All of this came from a definition 'by abstraction' of $l\text{'}u$, according to which this limit is introduced through its relations (=, >, <), but not in an explicit way. The attempt to define it in a *nominal* way *is equivalent to the identification between the segments of rationals and real numbers*, with which we shall have (*1899a*, 261):

$$l\text{'}u = \theta u; \qquad l\text{'}u \leq l\text{'}v . = . \theta u \supset \theta v; \qquad a \in R . \supset . a = l\text{'}\theta a,$$

i.e. it will already not be necessary to define Q in terms of upper limits: they have been *eliminated*.

Peano admitted that $l\text{'}u$ has many properties in common with the class θu; he even wrote: 'they only differ by the nomenclature' (*1901a*, 105), but he gave the pre-eminence to the ordinary language over the constructivist trend. The same happened with the possibility of identifying real numbers with segments of rationals: Peano regarded this way as feasible, though 'less practical' (*1899a*, 262). In defining the segments he also wrote that they are different from the reals only by the nomenclature (*ibid.*, 259). However he added that, although in some operations (e.g. +) they behave as such, they do not do so in others (e.g. if u is a segment, neither its reciprocal nor its square are segments). He therefore concludes that a real number is a 'distinct object' from a segment of rationals, as it can be seen through the following: $1 < \sqrt{2}$ and $\sqrt{2} < \sqrt{3}$, whereas if $\sqrt{2}$ and $\sqrt{3}$ are segments, then we should write: $1 \in \sqrt{2}$, but $\sqrt{2} \supset \sqrt{3}$. In short: 'The number 1 and the segment θ have different properties' (see also Couturat *1904a* and Kennedy *1974a*).

3.2.4. The 'logicist' arithmetic: Cantor

I now come to the most important point: I shall explain here Peano's ideas on what we could call *logicist arithmetic*. These ideas do not *develop* such an arithmetic, but at least they lead us towards it and pave the way for Russell's methodological view. However, for Peano himself it was mainly a logical possibility that finally had to be rejected for practical reasons.

The first definition of 'number' in logical terms was carried out through the concept of 'number of a class', already in the tradition of the pre-eminence of the cardinality of Cantor and Frege (*1891d*, §9). However, though it consisted only in the introduction of the finite cardinals, it was an important attempt since (i) it began a line that Peano was later suggesting as a possible future way; (ii) it affirmed that what is defined is only the attribution of an integer to a class. For $a \in K$ and $m \in N$, we have the following (*1891d*, 100):

$$num\ a = 0\ . = .\ a = \wedge;$$

$$num\ a = m\ . = \therefore a\ \text{-} = \wedge\ :\ x \in a\ .\ \supset_x\ .\ num\ (a\ \text{-}\ \iota\ x) = m\ \text{-}\ 1;$$

$$num\ a = 1\ . = \therefore a\ \text{-} = \wedge\ :\ x, y \in a\ .\ \supset_{x,y}\ .\ x = y.$$

However, though Peano afterwards offers several propositions about what he calls 'enumeration', including the operations + and < between the numbers of a class, the sum and the logical product of classes, and the univocal correspondence (Dedekind's *ahnliche Abbildung*, which Peano introduced already in 1889), it seems he does not yet realize that the equality between two numbers of a class can be expressed by means of a univocal correspondence. Perhaps this idea was found in Frege *1893a*, since in the review of this work Peano wrote (*1895c*, 194-5):

$$u, v \in K\ .\ f \in vfu\ .\ \bar{f} \in ufv\ .\ \supset\ .\ num\ u = num\ v,$$

as a transcription of an important proposition by Frege. However, though he resorts to Cantor, he only does so to mention that his concept of cardinal number was already expressed in the first *Formulaire*, but not to realize the basic idea of expressing the equality between cardinals as a correspondence, which had just appeared in Cantor *1895a*.

As soon as Peano knew this work by Cantor it seems he realized that its ideas, together with those by Frege, supposed the beginning of that 'logicist' view he was foreseeing, since it is from then on that in his works one finds more and more references to this possibility.[1] However it is not until *1897a* that we find this viewpoint explicitly linked to the idea of univocal correspondence (*1897a*, 216; *1897b*, 281; though he already clearly stated similar ideas in *1894b*, §26). The following are the main definitions (for $a, b \in K$): (i) the idea of correspondence (u) as a mere transformation (mapping) between classes ($a\ \bar{f}\ b$); (ii) the univocal correspondence (*Sim*, from 'similar'); (iii) the one-one correspondence (*rcp*, from 'reciprocal'); (iv) the 'number' of a class (*Num a*) (*1897b*, 236-7, 272 ff; *1897a*, 216-7):

---

[1] The fact that he wrote to Cantor asking for a definition of finite cardinal is very interesting. Cantor's reply, already implicit in his work, was: $1 = \overline{\overline{(a)}}$; $2 = \overline{\overline{(a, b)}}$ (see Kennedy *1980a*, 62, and Dauben *1979a*, 178).

$$u \in a \bar{f} b \ . = : \ x \in a \ . \supset_x . \ x u \in b;$$
$$u \in (a \bar{f} \ b) \ Sim \ . = : \ u \in a \bar{f} \ b : x, y \in a . x - = y \ . \supset_{x,y} . x u - = y u;$$
$$u \in (a \bar{f} \ b) \ rcp \ . = : \ u \in (a \bar{f} \ b) \ Sim : y \in b \ . \supset_y . \exists a \cap x \in (x u = y);$$
$$Num \ a = Num \ b \ . = \ . \ \exists (a \bar{f} \ b) \ rcp.$$

I have not yet seen the whole *1898d*, but in his first description of the philosophical implications of Peano's school achievements Russell wrote (*1900b*, 11, 13) that already in that work Peano had clearly pointed out (in a note concerning the proposition 210) that arithmetic could be developed *starting from purely logical definitions*, and that only this 'another way', by starting from a definition by abstraction of the cardinal number, makes transfinite numbers possible (which was decisive for Russell).

In *1901a* this new way appears with absolute clarity as well as the first steps on it (§32, pp. 70 ff). Starting from the previous definition by abstraction of the number of a class in terms of one-one correspondence, Peano says that since it is a definition expressed only in terms of the symbols of logic: 'Arithmetic can start from here', i.e. by logically defining 0, N_0, and +. Therefore we can construct arithmetic 'without going through the primitive ideas of §20' (which are precisely those of zero, number, and successor, that were introduced as *Pi* through the five axioms). Peano even points out the possibility of *eliminating* the definition by abstraction of the number of a class, by transforming it into a nominal definition in terms of a class of classes, although to reject it (*ibidem*):

> Given a class *a*, we can consider the class of classes: $Cls \cap x \ni [\exists (x f a) \ rcp]$; the equality of this *Cls* of *Cls*, calculated on classes *a* and *b*, involves the equality *Num a* = *Num b*; but we cannot identify *Num a* with the *Cls* of *Cls* considered, for these objects have different properties.

However, this rejection, which rests only on the mathematical usages and ordinary language (as we saw above and we shall see again in 3.4), would be strongly criticized by Russell in POM.

The definitions of the —up to then— *Pi* are carried out without problems, though giving rise to somewhat complicated formulas (see Jørgensen *1931a*, I, 190). Cantor's plan is always followed by identifying the number of a class in general (*Num'Cls*) with the N_0 for the finite classes, i.e. with Cantor's ordinals. Thus, 0 and 1 are defined by taking advantage of the ideas already expressed in 1891 (though now already in an explicit way), and finally finite numbers through the definition of infinite number (*infn*) (*1901a*, 71):

$$0 = Num \ \wedge;$$
$$1 = \iota Num \ ' \ [Cls \cap a \ni (\exists a : x, y \in a \ . \supset_{x,y} \ . \ x = y)];$$
$$infn = Num \ ' \ \{Cls \cap a \ni [\exists Cls \cap u \ni (u \supset a \ . \ u - = a \ . \ Num \ u = Num \ a)]\};$$

$$N_0 = (Num \ ` \ Cls) - infn.$$

This is followed by the definition of the usual operations and by the symbolization of many of Cantor's standard propositions (e.g. $Num \ (Cls \ 'N_o) > Num \ N_o$), although Peano recognized that 'the reduction of this theory into symbols is still very incomplete'.

This was the true challenge for Russell, who in fact already had undertaken the task that would lead him to PM through his first unpublished attempts and his work on relations. Peano's 'implicit logicism' (Aimonetto *1960a*, 605), I think, is undeniable, especially through the pre-eminence of the concept of class, the importance of nominal (explicit) definitions (see 3.4 below), and many 'logicist' statements spread throughout his writings.[1] Kennedy (*1980a*, 369-70) wrote that the logicist signs of *1901a* were the result of having read Russell's first article on relations (*1901a*). However, as we have just seen, the same way was open for Peano since much earlier through Cantor (and perhaps through Frege). In 4.3 I will show that probably Russell's definition of the number of a class in terms of a class of classes was taken from Peano, and in 4.5 that Peano's followers, especially Pieri and Burali-Forti, introduced the *full* logicist idea through the method to transform definitions by abstraction into nominal ones.

3.3. Geometry

3.3.1. The geometric calculus and the principles of geometry

Peano's second work (in 1888) was already devoted to 'geometric calculus'; but here I shall give his basic ideas through later works (*1891a* and *1896b*), and only in so far as they contribute to the understanding of his reductive methods. A geometric calculus was interesting for Peano because, following Leibniz, Moebius, Bellavitis, Grassmann and Hamilton, it made it possible to operate directly on geometric objects without resorting to the coordinates of analytic geometry (i.e. with no use of the numbers determining them), but through the application of the general rules of algebra.

The central idea of Peano's version of this calculus (proceeding mainly from Grassmann) is the possibility of reducing all geometry to four extremely general 'geometric forms' that, in certain conditions, can be expressed through sets of four points (tetrahedrons), and that make it possible

---

[1] For instance; (i) to assert that deductive logic is a part of mathematical science (*1888a*, 18) (the doctrine of the *fusion* from Boole, Peirce and Whitehead); (ii) to say that arithmetic, algebra and theory of numbers do not express sciences with definite boundaries, and that what is called logical order is only habit (*1898e*, 241). The clearer passage is already later to POM: 'il n'y a séparation ni opposition entre la Logique et la Mathématique, mais passage logique de l'une a l'autre' (quoted by Couturat *1904c*, 1046).

certain interpretations powerful enough to embrace the traditional concepts of point, line, surface and volume, along with vectorial calculus. Besides, extremely general relations and transformations between those forms are stated through various types of 'products', including analytic geometry as a particular instance and making vast applications in infinitesimal geometry possible.

The tetrahedrons of departure are sets (products) of four points (the vertices) for which magnitude, sense and equality are defined and, through a correspondence with the reals, the possibility of operating with those numbers. The geometric 'forms' are sums of points to which a real number has been assigned. The number of points of every element determines its degree; thus, x being the diverse numbers (coefficients) and A and B being points, '$x_1 A_1 B_1 + \ldots + x_r A_r B_r$,' will be a form of the second degree. Since the tetrahedrons are also products of points (of 4, 3 by 1, or 2 by 2), it will be sufficient to multiply every form by the difference between itself and four to obtain them.

Hence a point and a line will be particular instances of forms of the 1st and the 2nd degree (the line because it can be seen as a product of two points). Likewise, a triangle will be of the 3rd degree (through a *progressive* product). A vector will be of the 1st degree because, though it can be determined by *two* points, which are sufficient for fixing its origin, sense and length, however the latter is expressed as a difference, i.e. as a magnitude, which is the result of two elements each one of which having only *one* point. (This reduction of vectors to a more general concept was the advantage of Grassmann's method.) A bivector will be a form of the 2nd degree (like the pair of forces from mechanics), as the result of the (progressive product) of two vectors; and a trivector will be of the 3rd degree (already with no mechanical interpretation). The forms of the 4th degree are reduced by Peano to sums of tetrahedrons (reducible, again, to only one of them).

If we relate vectors to the common geometric figures we shall have: the (progressive) product of a vector by a point is a line; that of a bivector will be a triangle and that of a trivector a tetrahedron. As for coordinates, they will be, in general, the (numerical) coefficients of the elements of reference to which any form can be reduced. If the elements of reference are any form, the coordinates will be projective, but if, for example, they are a point and three vectors, we will have the Cartesian coordinates. The relation with projective geometry is narrow since the three first forms can determine, respectively, a projective point, a straight line at infinity, and a plane at infinity.

The *regressive* product determines the 'incidences' of the forms among themselves (and it is the historical result of the geometric principle of duality): a line (2nd degree) by a plane (3rd) is a point (1st); the product of two planes is a straight line and that of three planes is a point. Finally, the *internal* product of two vectors will be the product of their magnitudes by the cosine of the embraced angle, and the *external* product the corresponding bivector (equivalent to a vector perpendicular to the plane of the former ones, and whose magnitude is equal to the embraced area). With these concepts (also proceeding from Grassmann) Peano points out the possibility of giving an account of all the operations with complex properties.

In a parallel way to geometric calculus, and starting from Pasch's axiomatic method through the searching for *Pi* and *Pp*, Peano elaborated from *1889a* on another attempt to give an account of geometry (first projective, then metrical). I shall offer here only the essential ideas of the two basic works (*1889b* and *1894a*), which constitute the first *formal* axiom system for geometry.

1889b starts from three *Pi*: point (1), identity or coincidence (=) and the relation of internality ($c \in ab$: the point c is internal to —or is contained in— the segment ab). Afterwards there come the three traditional axioms about identity; other two axioms stating the figures or classes of points (*K1*) and the equality of segments ($a = b . c = d : \supset . ac = bd$); and a set of 14 definitions. Let us see those of ray ($a'b$), straight line (2), plane (3) and convex figure (*Cnv*) (*Cl* being 'collinear') (*1889b*, 61-2):

1. $a'b = : 1 . [x \in] (b \in ax)$
10. $2 = [x \in] (a, b \in 1 . a - = b . x = (ab)$" $: - =_{a,b} \wedge$
12. $3 = [x \in] (a, b, c \in 1 . a, b, c - \in Cl . x = (abc)$" $: - =_{a,b,c} \wedge$
14. $Cnv . = . [x \in] (x \in K1 : a, b \in x . \supset_{a,b} . ab \supset x)$.

From there on, and after 35 theorems deduced from the definitions, the 16 axioms of geometry 'of position' appear, each one of them (or each group) followed by their corresponding theorems. Here are some examples of axioms (*1889b*, 64, 66):

I. $1 - = \wedge$

II. $a \in 1 . \supset \therefore x \in 1 . x - = a : - =_x \wedge$

VII. $a, b \in 1 . a - = b : \supset . a'b - = \wedge$.

The first one states the existence of points; the second the existence of points not coinciding with a given one (repeated in XII and XV for the existence of points not contained, respectively, in a given line and a plane); the third the possibility of prolonging a segment through any of its points. Peano admits the coincidence with Pasch up to this point but, on not making use of Pasch's notion of 'portion of plane' (or *flat surface*, a limited plane), he claims the originality of what follows (which is like saying that he works with a *geometrical Pi* less). The axiom X states that if two points (c, d) lie on a ray ($a'b$), then either $c = d$, or $d \in bc$, or $c \in bd$, and the axiom XVI the tridimensionality of space. Finally, an appendix introduces one more axiom and discusses the concept of motion, but without introducing it.

1894a essentially repeats the same, but adds more comments, some improvements, and the explicit distinction between geometry of position and metrical geometry (which is also developed through axioms). The concept of motion is introduced, not through the relation of congruence (like

Pasch), but by means of the concept of *affinity* (*Aff*) (which rests on the 'logical' concept of correspondence, or mapping) (*1894a*, 143; *a*, *b* being points and *m* a mapping between points):

$$Aff = (pfp) \cap m \in [a, b \in p \;.\; c \in ab \;.\; \supset_{a,b,c} \;.\; mc \in (ma)(mb)],$$

i.e. *affinity* is a particular class of correspondences. Then motion (μ) is presented as a class of affinities ($\mu \supset Aff$), through a set of eight axioms containing identity as a motion ($\omega \in \mu$), as well as the inverse mapping ($m \in \mu \;.\; \supset \;.\; \overline{m} \in \mu$) and the product ($m, n \in \mu \;.\; \supset \;.\; mn \in \mu$). The rest of the axioms distinguish the class of motions from other affinities and they are presented as 'common cognitions' establishing the possibility of carrying out several translations. Finally, four types of motions are defined through a notation, inspired by logic ($_a^b$), where the lower elements are transformed into the higher ones (more details in Torretti *1978a*, 222-3).

In 1898 Peano wrote his 'Analisi della teoria dei vettori' (*1898c*) as a sort of fusion between the two previous parallel lines of work (geometric calculus and principles of geometry), attempting to apply, in geometry, the process that the year before had led him, in logic, to reduce the primitive basis to a minimum. The basic idea is now to reduce the theory of vectors to the smallest number of *Pi* (3) and *Pp* (16), starting, as in previous works, from the idea of point and from certain relations among points, and introducing the operations in the customary way, though dispensing with the concept of tetrahedron and arriving at metrical geometry through a third *Pi*: the internal product of two vectors.

The first two *Pi* are those of point (*pnt*) and relation of 'equidifference' among four points (i.e. $a - b = c - d$). This relation can be regarded like either equality of segments, or a parallelogram, or a translation, but it always leads to the notion of vector. The three first *Pp* introduce the traditional properties of equality (but now showing 'geometric facts') and two other state certain relations. Then the idea of vector appears (1898c, 191; $b - a$ being the difference between two points):

$$vtt = \overline{x \in} \; [\; \exists \,(a,b) \in (a, b \in pnt \;.\; x = b - a)],$$

and some definitions introducing the operations between vectors and between vectors and numbers.

The equality of forms of the 1st degree (or sums of points) is not already introduced through the concept of tetrahedron (a concept which 'has not yet been reduced to intuitive ideas'), but directly through vectors (which are particular instances of that concept): two forms are equal if, however 0 is chosen, the sum of the vectors (multiplied by their respective coefficients) going from 0 to the points of the first one, is equal to the corresponding sum of the second.

According to Peano, all this allows relations (and functions) of points to be expressed, but these relations are not altered in coming to affine figures, so that we cannot consider the concepts of distance, value of an angle, and metrical properties. That is why the *internal* product (u/v) of the two vectors is introduced as a new *Pi* (despite that in elementary geometry it is usually defined through the concepts that the idea of motion makes possible). Thus the need for analyzing the concept of 'cosine' is avoided (as well as those of 'length' or 'projection' if we define this product as the projection of a vector onto another). The new idea is determined by the four *Pp* stating: its result as a real number ($u/v \in q$); the commutative property; the distributive property as regards the sum; and the fact that the product of a vector by itself is a real positive number ($u/u \in Q$). Finally, the product of a vector by an irrational number is introduced through the concept of limit, several *Pp* about parallel and coplanar straight lines, and a *Pp* stating the tridimensionality of space. The later changes would only be minor ones.[1]

3.3.2. The 'logicist' geometry

Also in Peano's geometry, like in his arithmetic, we can find 'logicist' elements (at least in the sense that he made Russell's logicist geometry easier). We have already seen Peano's intention to present the geometric calculus as an axiom system starting from some *Pi*. This can be regarded as a logicist element since, though he recognizes a 'cut' in relation to arithmetic and logic (in introducing new irreducible geometric *Pi*), nevertheless he emphasizes the possibility of giving an account of all elementary geometry through the theory of vectors, which are presented in terms of completely abstract 'geometric forms' (in the style of Grassmann) and, therefore, very close to 'logical' forms. Besides, Peano insists that geometric calculus allows us to operate in exactly the same way as in algebra (with the differences noted above). Precisely to delve deep into this analogy he introduces the relation of equivalence in 1898, which gives rise to a fixed relation between points through a 'geometric' interpretation, and to an authentic calculus through an 'algebraic' interpretation.

Peano was perfectly aware of this idea when he wrote that, though they are subject to particular rules, 'algebraic, geometric and logical calculus have analogous operations' (*1894b*, 127). With this he not only wanted to emphasize the 'structural' similarity, but also the deep unity of a single science whose operations, regarded as functions, belong to logic (including notions as correspondence —mapping— and others which we now regard as belonging rather to set theory). Therefore he quotes several times with approval a passage by Tait (on Hamilton's quaternions) that, by clearly anticipating Whitehead's universal algebra, sets up that unique science and its several branches: 'there is no more than one single science in mathematical Analysis, which possesses diverse branches, but employing in each one of them the same procedures'. One of these branches will lead to the geometry of position (projective), which is deeper than ordinary geometry; the other to the deductive logic of Boole, which is analogous to the theory of quaternions. I think that Peano's attempts from 1898 on can be better understood as the result of the belief in this unique science. It is true that Peano (like Whitehead) did not choose logic as the

---

[1] In *1901a* two *Pp* were added. In later presentations the only change was the ordering and some notational improvements, though new definitions and concepts appeared.

ground for this science, but his whole view must have constituted a very attractive starting-point for Russell's logicism.

Within geometry there is still another 'logicist' element: the repeated attempt to introduce metrical geometry through a relation of correspondence (a mapping), i.e. in 'logical' terms (for Peano functions in general belong to logic). Already in an appendix to *1889b* he posed the possibility (that we saw before) of defining motion in terms of the already introduced objects, concluding that it can be reduced to the concept of correspondence or function (and thus 'to keep it as a part of logic') by defining 'homography' as a correspondence among the points in space, and the translation of a figure to another position as a particular homography. (The same that he did in *1894a*, as we saw, through the concept of 'affinity', a type of mapping with the advantage of allowing the introduction of various types of motion as 'logical' transformations.)

There is a final contribution, if not to logicism, at least to the attack against the power of the visual intuition: Peano's curve. Cantor had already demonstrated that a one-one correspondence can be established between the points on a line and those on a surface, which supposed a first landmark on the line of the loss of intuitiveness (an underlying constant in his work). Besides, he set the ground for the vanishing of the borderlines between algebra, geometry and logic (already implicit in Grassmann and Riemann), which would result in topology. However, the stated correspondence was necessarily discontinuous, as it was later demonstrated (see Kline *1972a*, 1016; Kennedy *1980a*, 31-2). Peano proved in 1890 that it is possible to establish a 'continuous' correspondence, in the sense that the points on the line correspond with those on the surface so that 'the image of the line is the entire surface and that the point on the surface is a continuous function of the point on the line' (*1890b*, 114). However this correspondence cannot be properly continuous (this had already been demonstrated as impossible) in the one-one sense, i.e. that the continuity involved at the same time the values of the two functions (the point on the line could not also be a continuous function of the point on the surface). Of course it was the celebrated curve filling an entire area by going through all its points.

The discovery contributed as well to showing that the concept of curve used up to this point was inadequate (see Kline *1972a*, 1016), but, from our point of view, the most important thing is that it stressed the non-distinction between line and surface. As the same was applicable to volume, it also contributed to erasing the difference between dimensions, which led to posing more and more serious doubts about the usefulness of the concept itself of space (and other analogous concepts as line, surface, solid) in geometry.

3.4. The method

3.4.1. Axiomatics

The most characteristic feature of Peano's method was axiomatics. Already in his first work of 1888 (*Calcolo geometrico*) the concept of vectorial space was for the first time axiomatically presented, by defining as 'linear systems' those systems of entities satisfying the expressed conditions (see Kennedy *1980a*, 23-4, and Bottazzini *1985a*). In *1889a* Peano was already completely aware he was making use of a 'new method', an expression that he then incorporated into the title of the work, whose preface explained that it was a method capable of being applied to any science and characterized by introducing initial symbols (named later *Pi*), postulates or axioms (*Pp*), definitions and, through demonstrations, a series of theorems. The process towards the smallest number of *Pi* began this same year in his work on geometry, where Peano already posed the problem of what geometric objects could be assumed without definition (and what properties without demonstration), and it can be said that it finished when logic itself was presented by him as an axiom system in 1897. The consequence was that only the *definitions* (i.e. certain identities) were capable of carrying out this reduction.

On the other hand, though the recognition that an infinite number of systems can satisfy a set of axioms, he soon arrived (1891) at accepting the possibility of presenting those sets almost as abstract structures. Thus, in offering, for example, the axioms of arithmetic as mere necessary and sufficient conditions so that the objects of a system (of an infinity of them) can be put in correspondence to numbers, he was really taking from numbers their traditional character of 'privileged' objects (except for the recourse chosen by Peano of considering them as 'what can be abstracted from all those systems'). This necessarily affected the status of definition and led him to the problem of whether it did not simply mean explaining properties previously given to intuition.

Consequently, it does not seem that Peano's axiomatics can doubtlessly be qualified as 'formal', if we mean not only to make use of a symbolism, but to emphasize the 'open' character of the *Pi*. However, Peano's intention of demonstrating the independence of the axioms clearly shows that it was the interchangeability that appeared for many interpretations which pointed out the existence of an underlying 'formal' structure (see Torretti *1978a*, 231). This prevents us from regarding his axiomatics as 'intuitive', i.e. like that of Euclid (see Aimonetto *1963a*, 602). As for logicism, according to what we have seen, it was not far from this relation between intuition and analysis: it is characterized mainly by 'constructing', through a chain of definitions, all mathematical objects out of a few logical ones intuitively known. Dedekind 'extracted' his axioms from an analysis of the series of natural numbers and he thought, nevertheless, that he reduced them into logical notions; and Peano himself knew, as we have seen, the way to build a logicist arithmetic starting from a method to logically 'define' numbers. Here I shall try to delve into all these concepts just as Peano himself used and interpreted them.

Concerning *demonstrations* Peano was always aware of his logical nature; with this, he reached a position, together with B. Peirce and Frege himself, next to whom defended that mathematics draws 'necessary consequences'. This view on demonstrations led Peano to describe them as 'transformations' of the *Pp* until obtaining other derivative propositions; likewise in algebra, the step of a proposition to the following would be an identity (*1889b*, 81). The problem would of course be to find the basic propositions through intuition (*1889b*, 82), i.e. we would always want to know whether it is possible to demonstrate a particular proposition, which will lead us to searching for other propositions from which this can be derived. That is to say, on the one hand we must try to recognize what propositions 'express the simplest properties of the regarded entities' (*ibid.*, preface); on the other hand, if demonstrating means only deriving through a transformation, the relativity appears (since any transformation is always referred to the set of propositions taken as *Pp*). Therefore, to know whether a proposition can be demonstrated will depend on the particular set of propositions we have chosen as primitive (as happened with definability) (*1891c*, 103-4).

The solution will be the 'empirical' comparison: if we start from a particular proposition, to know 'whether it expresses a truth in the simplest possible way results only from its comparison with others' (*1891c*, 104); i.e. we shall regard as *Pp* those propositions that 'cannot' be deduced from other simpler ones. In practice it supposes not being able to leave ordinary language: we shall manage to find the rules to be followed only by analyzing the well formed reasonings that are actually carried out in any field where the logical rigour is intended (*1894b*, 174).

In this way we also come to relativity: when Peano tried to reduce 'simple' rules to the smallest number he had to recognize that the selection depended upon intuition, since several demonstrations can be found for a same proposition, and sometimes two propositions are reciprocally derivable (as happened in Russell's manuscripts; see 2.2 and 2.4 above). It gave rise to the concept of 'possible demonstration', a concept parallel to that of possible definition. Let us see now the way in that the notion of 'independence' contributed to the problem.

It was Boole who (in his *1847a*) related the idea of interpretation to the 'structural' solidity of a system, although Grassmann already made it explicit the idea that his 'forms' maintained the same internal relations through different meanings. Pasch as well regarded the possible diversity of meanings as a virtue and a help to a sound system of reasoning; and for Dedekind the proof of the validity of a system consists in seeing whether it holds in changing the meanings of the involved terms. The next step was to associate the independence of an axiom, in relation to a system, to the possibility of offering an interpretation, instance or model, satisfying all the axioms of the system except this one; and this step was an achievement due to Peano (although there is the general precedent of the non-Euclidean geometries, that can be regarded as interpretations proving the independence of the axioms of parallels).

Peano himself was gradually closing in on the idea. In the first works he only points out that he cannot (or does not know how) to deduce some axioms from the others, which should have been sufficient to be convinced of their independence (*1889b*, 57). However, in practice, he offered several samples of interpretations to show the way in which diverse relations hold in them

(*1889b*, 77, 79, 83, 84, 90). In *1891d* (pp. 87-8) he made a systematic use of the method by offering examples 'satisfying' every one of the five arithmetical axioms except that whose independence has to be 'seen'. With that it seems that Peano arrived at the method by gradually associating the capacity of finding interpretations with the structural solidity.

Only in *1894a* (p. 127) was the method theoretically described: to prove the independence of some postulates it suffices with obtaining examples by attributing any meanings to the undefined terms, in a way that they satisfy a set of propositions but not all of them, thus we have proved that the latter are not a logical consequence of the former. However he still explains the idea in terms of simplicity: he is not yet clearly aware of the concept of 'possible demonstration': once proved the independence, it can only be affirmed 'that we do not know how to analyze the propositions assumed as postulates into other simpler ones'.

Peano made later use of the method mainly to show the persistence of particular logical laws, but, before the discovery of 1897 about the multiple relations of interdefinability among the *Pi*, he insists on the customary intuition. This is the reason why since *1898c* (pp. 188-9) he admits that certain ideas are 'assumed' as primitive without interpretation, with the argument that we can assume those properties as primitive on which the rest of the properties depend. But it is not explained *how we can know which are the properties of an idea without interpreting it*; a position that, in geometry, leads him to hold that the propositions express *facts* and that what is intuitive is what is simple (*1898c*, 198), and, in arithmetic, to hold that there is a way to abstract what all systems satisfying the *Pp* have in common (*1898d*, 216-8).

Peano would still introduce two novelties to this subject. The first one was the distinction between ordered independence (in relation to the *preceding* axioms), and *absolute* independence (referring to *all* the axioms of a system), although for that he already mentions his collaborators Pieri and Padoa (*1900a*, 322; *1901a*, 43-4). The second stated the relation between independence and consistency and, curiously, appeared in the Paris Congress of Mathematics in 1900 in response to the celebrated problem no. 2 by Hilbert (see the details in Kennedy *1980a*, 97). In 1906 Peano improved his explanation: if the axioms of arithmetic are all fulfilled under certain interpretation, it is proved that they are necessary and sufficient and that there is no contradiction among them (*1906a*, 343; see Pieri *1906a*). In other words: Peano spoke of what today is sometimes called semantic consistency, although he really continued to depend on the intuition in adding that these proofs of consistency are useless since the axioms of arithmetic are 'simple'.

3.4.2. Definitions

The constructive character of definitions was systematically searched for by Peano, who resorted to other types only when he could not offer nominal definitions of the usual type, i.e. those in which the defined entity was the result of a combination of other entities already introduced; but as well in those other occasions the results were interpreted within the reductive framework. We can describe the evolution of this subject in four stages.

(1) The first one (which culminated in 1894 with the first systematization) already stated the basic characteristic: a definition is a proposition of the form '$x = a$' (sometimes preceded by some condition), where 'a' is an aggregate of symbols possessing a known sense and 'x' a symbol (or aggregate of them) still lacking significance (*1889a*, xvi). The consequences were quickly drawn (*1898b*, 56, 78), mainly as regards the need for asking for the ideas that cannot be defined out of a given set of them. This leads us to the well-known arbitrariness, especially when we intend to reduce the primitive ideas to the smallest number and the interdefinability appears. In any case for Peano it is sufficient with the remark that mathematical definitions always belong to this type (*1899b*, 78). Other consequences are the following: (i) definitions are nominal and serve as abbreviation, except for those that are only explanations (e.g. those by Euclid); therefore, we cannot speak *in general* of the definability of a concept (*1891c*, 103); (ii) to define, from this viewpoint, cannot be anything but 'to construct' (Peano does not make use of this term), i.e. to form an expression identical to the *definiendum* out of *Pi* or ideas previously defined (*1894a*, 116); (iii) it can also be said that the axioms 'define' the *Pi*, in determining their properties; this makes the constructive character implicit, but it also contributes to make axioms and definitions closer (*1894b*, 175).

All that leads to the first list of rules for definition (*1894b*, §41). The most relevant ones for us are the following. Definitions are not necessary (they can be replaced by what they stand for); this is why they do not have to be proved: they are a result of our will to abbreviate. Nor does the existence of what is defined have to be proved: it is possible to define non-existing 'entities'. So Peano breaks an entire tradition, but he also contributes to the constructive methods (in fact he used to introduce 'existence' through axioms). The fact that not everything can be defined leads, according to Peano (and Russell; both in the Aristotle-Locke tradition), to admit 'indefinable' ideas. However, *though he identifies what is undefined with what is simple, he now holds that simplicity is arbitrary*, i.e. relative (*1894b*, 173-4). Though he still needs to relate simplicity to language, he does not yet do it with ordinary language, but with the language of science.[1]

(2) The second crucial stage is the controversy with Frege. Peano claimed especially for the possibility of offering different definitions of the same symbols through the recourse of completing, with previous hypotheses, the meaning of the variables (see my *1988b* for details). With this he thought to be more consistent with the practical needs of mathematics and its many extensions of concepts and operations, but Frege's criticisms and his accusation of ambiguity led him to an attempt to reduce the ideas of logic to the minimum and also to offer 'general' definitions (e.g. that of identity) which, in some way, established a primary meaning which other secondary ones could be referred to. However already before undertaking this attempt he realized that he was handling various types of definition: (i) that of a symbol, without hypothesis; (ii) that of an expression, always with some hypothesis; (iii) definition 'by abstraction'; (iv) 'implicit' definition.

---

[1] The condition of homogeneity was also introduced at this stage: the same real variables must be contained in both sides of a definition (see Jourdain *1910a*, II, 302, where a good description of the evolution of this condition can be found).

(3) 1897 is another important point. The main result of Peano's systematization of logic (from our viewpoint) is the possibility of interdefining the fundamental symbols, which leads us to the relativity of any logical order, i. e. the 'possible definitions' (*1897b*, 221); to consider the reciprocal reducibility as the source of the arbitrariness of every selection of *Pi*, and, finally, to the primacy of 'convenience' as a criterion for preserving some symbols and eliminating others (*1898a*, 287). Thus, leaving aside the definitions by induction (already in Grassmann; see Wang *1957a*) and the implicit definitions, there remain three types: (i) those stating abbreviations or combinatorial possibilities, but with no elimination; (ii) those introducing new symbols as a construction out of those previously introduced (which, therefore, involve a reduction); (iii) those stating other possible logical orders. This seems to imply, for Peano, a new valuation of ordinary language, which comes to acquire a greater importance as a selective criterion. In such a way, the *Pi* are led to be interpreted as a sort of 'minimum vocabulary': to attribute a value to the definitions of terms of ordinary language it is necessary that first a 'table of [known] words' is built, to be used as the only admissible material in the constructions, whose final result will be to reconstruct the logic itself (*1898d*, 243-4).

(4) In 1900 the process arrives at its end: while in 1897 some reductions were regarded as real definitions (*Df*), now the majority of them are mere possible ones (*Df?*). On the other hand, certain definitions appear which are accompanied by the symbol they are defining (e.g. *Df* ∩), as announcing that this definitions will be adopted as 'official' among other possible ones (*1901a*). In this way Peano seems to be interested in clarifying the several types of definitions that were mixed in 1897. (However, it could also be interpreted —like Russell in POM— as if he was giving up the attempt to emphasize a certain number of *Pi*.) Now the *Df?* are presented as *intrinsic* and the *Df* as *relative* to a given logical order (*1900a*, 319), with which a change in this order will turn certain *Df* into *Df?* In practice, in *1901a* the symbol *Df* will only be used (in general) to those definitions combining symbols, without eliminating them from the *definiens,* and the symbol *Dfp* (the same as *Df?*) to those reducing a symbol into others (which leads him to emphasize certain reductions as privileged). Here are some examples (1901a, 11, 19; *a, b* being classes):

$$x, y \in a = x \in a \:.\: y \in a \quad Df;$$
$$a \supset b \:.=:\: x \in a \:.\: \supset_x \:.\: x \in b \quad Dfp;$$
$$a \cup b = x \ni (c \in Cls \:.\: a \supset c \:.\: b \supset c \:.\: \supset_c \:.\: x \in c) \quad Df.$$

The final stage is the culmination of the process of technical conventionality and pre-eminence of ordinary language, as it is shown by the explicit admission that definitions are true only by convention: they are either historical facts or a result from the will of every author (*1901a*, 7). *1900b* (like *1911b* and *1921a*) adds almost nothing in spite of being fully devoted to this subject: it is limited to gathering the ideas from preceding works.

3.4.3. The definition by abstraction

The relation between equivalence and abstraction appears very early in Peano and, of course, is prior to the expression 'definition by abstraction' itself. In *Calcolo geometrico* (1888) he regarded abstraction rather as a process of equality among functions, although he decided later (together with Burali-Forti) to introduce it as a new type of definition (something similar to what happened with the 'definition by postulates'). The first reference to this relation (which already summarizes the essentials) starts, however, from the concept of relation; in particular from a relation of equivalence. From there on it is stated that any equivalence not being an identity can be regarded as an identity among certain entities, obtained by abstracting (from —and only— all of these entities) the properties that distinguish an entity from those being equal to it. Thus, 'A is parallel to B' can be written as 'the direction of A is identical to that of B' (see Jourdain *1910a*, II, 272).

The link with functions (and correspondences) appears when the concept of 'presign' of a function into a class is defined (*1889a*, xiii), and the only novelties prior to the first systematization of the definitions by abstraction are already only to call 'equivalence' among propositions to the mutual implication (*1891b*, 92; *1891c*, 108) and to regard fractions and real numbers as introduced 'by abstraction' (*1891d*, §§10-11).

In *1894b* (pp. 167-71) the expression 'definition by abstraction' is introduced for the first time. The *locus classicus* is the following passage (translated in Kennedy *1974a*, 397, with some changes by myself):

> Let u be an object; by abstraction, one deduces a new object φu. We cannot form an equality
>
> $$\varphi u = \text{known expression},$$
>
> for φu is an object of a nature different from all those that we have considered up to the present. Rather, we define the equality $\varphi u = \varphi v$ by setting
>
> $$h_{u,v} . \supset : \varphi u = \varphi v . = . p_{u,v}$$
>
> Where $h_{u,v}$ is the hypothesis on the objects u and v. Thus $\varphi u = \varphi v$, being the equality defined, means the same as $p_{u,v}$ which is a condition, or relation, between u and v, having a previously known meaning. This relation must satisfy the three conditions ...

The advantage of that takes place in those cases where φ is the symbol that has to be 'defined'. Thus, the new object φu is introduced only through equality and it is, therefore, what is obtained considering in u all (and only) the properties common to the other objects v such that $\varphi u = \varphi v$.

Peano offers many examples; here are some of them: (i) the ratio of two magnitudes as equality of two ratios (Euclid); (ii) the rationals through $b/a = d/c . = . ad = bc$; (iii) the reals through the equality of upper limits (see 3.2.3 above); (iv) the length through congruence (or equality of lengths); (v) the parallelism through the equality of directions; (vi) the concept of vector through the equality between two pairs of points, etc. The subsequent treatments offer no novelties except for the stressing of the fact that such definitions carry out the *reduction* of a relation to an equality between functions (or abstractions) of the considered objects (*1901a*, 17).

The three basic examples of this kind of definitions in *1901a* (§§32, 61, 91) take place precisely in the peak moments of the chain of definitions constituting the work: when cardinals, reals and vectors are introduced, respectively founding arithmetic, analysis and geometry. However, the value of each one of them is unequal, given that only in the second case does Peano rest in fact on a definition by abstraction to continue the chain of definitions (see 3.2.4 above). On the other hand, number is introduced as a *Pi* and the equality of numbers is emphasized *precisely* to indicate the possible way to a 'logicist' arithmetic and the introduction of Cantor's transfinites, all of which are based on the correspondence or non-correspondence between classes (see 3.2.4 above). As for vectors, we have seen (in 3.3.1) that, though they are defined, the definition is based on the previous introduction of a *primitive* relation: the equality between pairs of points, which is determined by the corresponding *Pp*. Russell's task in POM (as we shall see in the next chapter) was to overcome these three obstacles and to achieve that this chain of definitions becomes continuous, i.e. without the introduction of new *Pi* (or gaps), which was only possible through the transformation of definitions by abstraction into nominal definitions, the only really *constructive* ones. (However, the means for this transformation were also provided by Peano and his school; see 3.5.2 below).

Peano, in general, thought that nominal definitions are the most important ones, but he tended to reject explicit logicism. If we remember that, since 1898 Peano forgot the construction of an 'ideal language' that could give an account of the multiple senses of the logico-mathematical terms, we realize that such a rejection can only be referred to the need for respecting ordinary language and the established usages. For this reason, the constructive definitions, no matter how many 'logical' virtues they have, should not abandon the guidance of this usual language. Some years later, Peano referred to this more clearly: in *1913a* (p. 403) he gives the general method for turning a definition by abstraction into a nominal one (following Pieri and Burali-Forti) in a way that, for instance, the reals become segments of rationals and the number of a class becomes a class of classes in one-one correspondence to that one. However Peano concludes: 'Although I do not deny the possibility of doing so, I deny the convenience. These nominal definitions, apart from attributing to the defined object ... the properties it actually has, *they also attribute other properties to it, which it does not have according to ordinary language*' (the emphasis is mine; the same in *1915b*). In 3.5.2 we shall return to this subject in more detail.

3.4.4. Simplicity, analysis and intuition

I shall try here to clarify the relation of the concepts from the title (of this section) to the concept of definition, in order to state some ideas, although without trying to supply a systematic exposition that cannot be found in Peano's writings. Peano was not a 'formalist' (nor a 'symbolist') as he always intended that his symbols stand for intuitive ideas (Cassina *1933a*, 487) of ordinary language. However, as soon as one tries to reconstruct his theory on this subject, one finds dozens of scattered passages hardly compatible among themselves. Nevertheless, it seems we could speak about three stages.

(1) 1889-1891. Here Peano sets the foundations of what would be his method, through a first concept of simplicity as a synonymous to conceptual economy, according to which he attempts to find the definitions and demonstrations starting from the simplest ideas and propositions, i.e. a method that 'analyzes' the propositions by setting with precision the 'value' of the involved concepts (their meaning). The possible arbitrariness of the selection is then palliated by this simplicity. However, the primitive propositions of arithmetic already suppose an analysis of number as something 'intuitive', and the fact that a chain of definitions coming from these *Pp* can be built, creates a link between simplicity and definability. The essay on the concept of number constitutes the official claim from which the *Formulaire* will try to expand the method to all mathematics, in searching for the analysis of complex ideas into their simple parts. However in this way a more atomistic than economical concept of simplicity already appears (which affects the *Pp* in the same way). The fact that Peano regarded the *Pp* of arithmetic as an analysis of number should not make us think that for him definition and analysis are different methods, since the axioms also 'define' the concepts by determining their properties. Finally, the access to such simple properties has to be 'intuitive', since they are indemonstrable; it is what Peano expresses by saying that they have to be obtained by induction (abstraction) and not by deduction.

(2) 1894-1896. The first wide systematization of logic shows this science as a true instrument of reducing theories to symbols, i.e. of making a true analysis into simple parts possible. Firstly through the replacement of the 'logical terms', then through the analysis of the ideas of the concerned science. However, the status of simplicity comes to be arbitrary as soon as diverse alternatives are possible, which also concerns simple ideas. Peano tries to avoid the conclusion by declaring them as something acquired through the experience (the *induction* of the previous stage), and then as common cognitions to all human beings (*1894b*, 173-4). This trend is stressed in geometry, where it is openly claimed that the axioms must proceed from observation and they are 'implicit' in the empirical reality (*1894a*, 116, 141-2). Here, like in Pasch, the access to the first foundations is said to be 'intuitive', just to subsequently erase this character on turning such *Pi* and *Pp* into mere symbols. However, perhaps for this same reason, the conviction that under the different interpretations underlies a *unique and objective analysis* becomes more and more valid (*1895c*, 190; *1896a*, 198).

(3) From 1897 on, a 'complete' analysis of the ideas of logic is provided (*1897a*, 201). Nevertheless, the paradoxical interdefinability of the ideas that can be found in this way leads Peano to see that the recognition of what is primitive is, in logic, much more difficult than in arithmetic or geometry (*1897a*, 204). That is why his language on the logical *Pi* shows no precision, and though he describes them as 'the simplest ones', he refers to them only through examples and always recognizing their conventionality. On the other hand, in geometry the trend of considering the *Pp* as intuitive is stressed, until identifying what is intuitive and what is simple (*1898c*, 198). The same occurs in arithmetic, where the usual ideas of number and successor are

declared simple, in spite of admitting that an infinity of systems satisfy the *Pp* which determine them (*1909d*, 216-8). Concerning definition, Peano continues to insist on its identification with analysis: stating a chain of definitions is the same as analyzing the corresponding ideas. The belief in a unique analysis is also defended here (*1898e*, 242-4), at least if one admits that the five primitive propositions constitute a sort of definition. Finally, in a passage from *1906a* (p. 343), Peano, as it happened in other fields of his work, is less cautious with regard to philosophy and claimed that the axioms of arithmetic (like the ones of geometry) are empirical (being linguistic), and, however, objective.

This amalgam of psychologism, Platonism, reductionism and philosophical analysis had to be clarified by Russell, although doubtlessly it already contained the resources for which he was then looking.

3.5. Peano's followers and their contributions

I have very little room to mention the people, mostly Italian mathematicians, who were fascinated by Peano's original methods and his strong personality. Kennedy *1980a* has offered a list of its components and a minimal description of their work. About his characteristics as a 'school', Mangione *1972a* has described its negative features as a closed, local, and too dependent on the 'master' movement. Here I shall only briefly mention some of the improvements introduced into Peano's system by his followers, and then I shall explain what, from our viewpoint, was their major achievement and doubtlessly the more influential on Russell's logicism: the creation of a method of transforming definitions by abstraction into nominal ones.

3.5.1. The various improvements

In logic, the improvements by the followers mainly affected the details, except for Padoa's application of the method to prove the independence of the axioms to the *Pi*, and his attempts to clarify the essential characteristics of any deductive theory, to which Pieri also made some contributions (I shall refer to all this later). The majority of these improvements simply proposed certain changes in the notation to make the system clearer or to avoid complications. For instance, the symbol '$=_{df}$' from Burali-Forti ('$\equiv$' in Padoa), similar to the one adopted by Peano (= ... *Df*), all of them tending to avoid the implicit circle of defining, for example, the sign of equality, and, in general, to emphasize the importance of definitions in relation to other kinds of propositions. Another improvement was the introduction of the idea of the unit class through the symbol '*Elm*' by Padoa, who in his *1899a* also proposed the modification consisting in the recognition that everything on the right side of the symbol of membership is a class, which supposed the introduction of a new axiom in the arithmetic of *1901a*.

In arithmetic I shall mention two contributions. Firstly the one from Burali-Forti, who in a series of articles (mentioned later) tried for years to *reduce* the ideas supposedly indefinable by Peano to their logical constituents. This was always accompanied by an attempt to reduce all the algebraic operations to their formal properties until achieving a didactically more operative framework (especially in his *1899a*). Secondly we have Padoa's work, who in *1900a* offered an algebraic theory of integers, characterized by only three *Pi* and seven *Pp*, which went beyond what had been achieved until then as he systematically applied his method to prove the independence of the undefined symbols, and improved Peano's ideas on the independence of the axioms. The essence of this work was summed up in his *1900b*, especially by emphasizing the demonstrations of compatibility and irreducibility. Both works[1] were presented in Paris to the Congresses of Philosophy and Mathematics in 1900, with Russell's attendance (at the first one).

In geometry we have to mention especially Pieri (to whom I shall refer below in 3.5.2) and his achievements in the construction of genuine axiom systems, for general geometry as well as for projective geometry. With that, he continued the tradition of the followers, consisting in improving upon each one of Peano's innovations and applications, which (with regard to geometry) had started shortly after the appearance of the first relevant work by Peano in 1889, for example through the attempt by Vailati *1892a* to reduce the number of axioms. The two most important achievements by Pieri are *1898a* and *1899a*, whose method and basic resources were also presented to the Congress of 1900 through his *1900a*. I shall say something about each one of them in what now follows.

Pieri *1898a* is the culmination of an entire series of attempts (summed up in Couturat *1904a*, VI, 818) to apply the axiomatic method to projective geometry,[2] improving Peano's works. (For us it is also interesting because it was the only work from Peano's school that Russell knew about before his attendance at the Congress. His offprint contains underlined parts and notes which allow us to have an idea of the most interesting things for Russell at that time.[3]) The work reduces projective geometry to two *Pi* (the projective point and the joining-line [*congiungente*] of two

---

[1] Bowne *1966a* offers a good (although somewhat superficial) summary of the articles that Peano's school submitted to the Congress (of Philosophy) in 1900.

[2] See Couturat *1904a* (VI, 818 ff), where such attempts are reviewed and a good summary of this article by Pieri is offered.

[3] This fact is indicated in Grattan-Guinness *1986a* (p. 107), and it was confirmed to me by Kenneth Blackwell, who supplied me with a photocopy. In its title-page we can read: 'B. Russell. March 1898' in Russell's hand. We shall see the significance of the notes and emphasized passages by Russell mainly in 3.5.2 below. Of course, it can be objected that Russell could read the article *after* being familiar with Peano system, but I don't think so: first, the paper can be easily read with no special knowledge of Peano's notations (it is written in Italian, with only a bit of this notation, which is accurately explained); second, it would explain the great change towards a *formal* viewpoint in geometry in Russell's second reply to Poincaré in 1899 (see my *1990a*). However, it is not *known* whether Russell read Pieri's work in 1898.

projective points) and 19 axioms (*Pp*), all this in the framework of what tries to be a purely hypothetical science, i.e. a science independent from intuition.

After the introduction of the two *Pi*, which are claimed as having an arbitrary content (being determined only by the *Pp*), diverse axioms appear. The first two are: (I) 'The projective point is a class'; (II) 'There is, at least, one projective point'. (The work is written in Italian, but Pieri points out in the introduction that the original investigation was developed through Peano's symbolical notation.) From that, the various complex entities and the axioms stating the relevant entities are introduced: axiom III introduces the existence of one more point; axioms IV-X introduce and develop the straight line as a class of projective points, through the second *Pi* (the joining-line); axiom XI allows coming out of the straight line through the existence of a point outside of it (by previously defining collinearity, which makes it possible to introduce the *projective straight line* as a class of classes of points); axiom XII introduces the *projective plane*; from that, the rest of the concepts needed in projective geometry are developed: the construction of von Staudt's quadrilateral, the projective segment, etc., until arriving at the axiom of continuity (XVIII) and at the one stating the three dimensions (XIX).

In spite of the obvious axiomatic structure, the entire work shows a logicist look; not only because Pieri explicitly claims, like Peano's previous attempts, that projective geometry is only a logical combination of the *Pp* and a complex construction of the *Pi* (p. 4), but also because, although he admits proper geometric *Pi*, he hints that in the last analysis all the notions he makes use of are reduced to pure logical concepts.[1] Thus, he clearly claims that just as in algebra, where one starts from the ideas of number and function, the same occurs in geometry with the concepts of figure (point, straight line, etc.) and transformation, and he adds that with these 'we do nothing but to reproduce aspects or special modifications to two fundamental logical categories: *class* and representation [mapping]' (p. 43). As we shall see in 4.5.2 this is closely related to his interest in offering only *nominal definitions* (p. 3). On this same line, *ordinary space* is defined at the end as 'the class of all possible interpretations in the domain of those postulates [axioms I-XIX]' (p. 56), clearly anticipating Russell's parallel definition in POM.

One final point of interest is the novel introduction of a section (§7) devoted to *order*, which emphasizes just some notions that were fundamental for Russell at this time. The first thing to describe is the fact itself of granting importance to order, isolating it as a basic concept, especially if we take into account that it is defined as a transformation that at every point on a straight line assigns the points represented by a certain function, and it is identified with the notion of *projective motion*. The second important thing is the terminology and notation for *relations* that, doubtlessly, had to impress Russell just at a moment when he was very interested in developing a classification of them (see 2.6 above). For example, Pieri demonstrates a theorem exhibiting the incompatibility of two different orderings, which Russell describes at the margin (on his copy) as 'Sequence asymmetrical' (p. 33); further on the author offers another theorem stating the fact that given a straight line r with a particular ordering a, b, c; if x and y are two points on r, then all the points

---

[1] I suppose that is why Torretti (*1978a*, 229-30) distinguishes Pieri's goals from Hilbert's and emphasizes that Pieri was more interested in *reducing* the number of *Pi* than in improving an axiom system in itself.

following y will also follow x: $\sigma_{a,b,c}\, y \supset \sigma_{a,b,c}\, x$, i.e. by emphasizing that the first 'figure' contains the second. This is exactly the same as Russell's later notation for transitive relations (e.g. in *1901a*, where they appear as $R^2 \supset R$). Russell himself shows his interest on this in writing at the margin: 'Sequence transitive' (p. 34), just like on the next page, where Pieri offers an analysis of the possibilities of the ordering of several points on a straight line.

In *1899a* Pieri developed a similar system for elementary geometry, where this is reduced to only two *Pi*: point and motion (p. 175). This new attempt is also interesting because there Pieri is forced to *define* the straight line through the notion of joining-line (*congiungente*), which, again, is introduced as a class of points not being affected by certain *motion*. It was a fundamental idea for POM since in that work Russell made the effort to offer a logical definition of 'straight line' to achieve the *elimination* of any proper geometric *Pi* (see 4.4 below). Russell himself recognized this doubt (POM, §395), as he had already done so by mentioning Pieri *1898a*, in the chapter on projective geometry of POM, in describing it as the best work on this subject (POM, §363).

1900a is the work submitted to the Paris Congress. Pieri did not add new ideas to his former works, but he made the logicist idea to explicitly underly all of them (including the title pointing out geometry as a purely logical system). I have no room to study the content in detail; it will be sufficient to say that he emphasizes the fact that geometry is nothing but the study of 'a certain ordering of logical relations' (p. 368), and he recognizes that arithmetic itself is 'a part of Logic' (p. 370) (like Dedekind, although for Pieri 'logic' meant a different thing). Finally, a system of geometry is included as an example of the methodological notions (which we shall consider below).

I come now to the improvements that Peano's followers introduced in the *methods* in themselves. Here we are forced to emphasize again the writings by Pieri and Padoa. Pieri carried out a constant effort to clarify the essence of what he called 'logical (or hypothetical) deductive systems'. Already in his *1898a* he pointed out the purely deductive character of projective geometry, an 'abstract' science being completely independent from any physical or mathematical content, which can be transformed, through suitable interpretations, into elementary and metrical geometry. As such a deductive science, it remains independent of any intuition and, by means of the introduction of certain *Pi* and *Pp*, it develops all its content. This vision leads Pieri to present the *Pi* in a completely modern way (even before Hilbert, who never mentioned him), following Pasch (to whom he refers): the *Pi* posses an 'ideal content' denoting any primitive object which is determined only by the *Pp* (p. 6). This would contribute to explain, incidentally, the ideas sketched by Russell in his reply (*1899b*) to Poincaré, in describing geometry as a formal calculus (like Pieri in his introduction to *1898a*).[1]

This key idea was developed in Pieri *1900a* where, in a very wide historical framework, geometry is presented as a science independent from the possible interpretations of its *Pi* (p. 376), and these ideas are presented as *implicitly defined* by the *Pp* (p. 378), even by explaining the concept of implicit definition as a 'logical description' (p. 387), in the tradition of Gergonne

---

[1] Therefore Torretti (*1978a*, 307) does not seem right when attributing this originality to Russell (see however note 3, page 128 above).

1818a, who is not mentioned. Finally Pieri also mentions the possibility of proving the consistency of a set of *Pp* by means of *a sole example* verifying all of them at the same time (p. 380). In fact already in his *1898a* (pp. 60-1) he had devoted an appendix to this subject where he explains Peano's ideas on logical independence and where he introduces the idea of *ordinal independence* (it seems that anticipating Peano; see 3.4.1 above) and the method of proving the consistency (*1906a* was an article devoted to this subject).

As for Padoa, he devoted pages 314-24 of his *1900a* to clarifying some features of 'deductive theories' and to the introduction of his method to prove the irreducibility of the *Pi* (by applying Peano's general idea for axioms). We must emphasize his insistence on the notion of 'symbolical definition' (the definition from Peano) and on the need for proving the *existence* and *uniqueness* of the individuals to be defined (previously to give their definitions) (p. 324). Concerning irreducibility, this is the clearest passage (p. 322):

> To demonstrate that the system of undefined symbols is irreducible in relation to the system of undemonstrated *P*[propositions], it is necessary and sufficient to find, for every undefined symbol, an interpretation of the system of undefined symbols which verifies the system of undemonstrated *P* and continues to verify it if one changes the meaning of the considered symbol.

Given the great transcendence of this recourse for the development of the foundations of axiomatics, it is hardly necessary to add any comment. In any case Russell was never interested in this kind of metatheoretical investigations; this can be interpreted as a sign that his main interest in logic was to find and to apply certain instruments which are the proper features of a *philosophical* general method: that of constructive definitions.

3.5.2. The transformation of definitions by abstraction into nominal ones

I shall finish the chapter by explaining the genesis of the most important recourse that made Russell's logicism possible. We have already studied in detail the contribution of Peano to the concept of 'definition by abstraction' and its possible conversion into nominal ones (see 3.4.3 above). Now we shall verify that the essential idea on this subject was originally introduced by Pieri and applied for the first time to arithmetic by Burali-Forti.

Peano introduced definitions by abstraction in his *1894b* (see 3.4.3 above). However, in the same year Burali-Forti published his *Logica matematica* (it seems *after* Peano *1894b* according to Kennedy *1980a*, 53). In this work he presented a classification of definitions into four main types: by abstraction, by induction, by postulates and nominal. Only the last type was regarded as satisfactory (see Peano *1899a*, 261). With that, Burali-Forti anticipated his view in the Paris Congress in 1900. Burali-Forti *1900a* may be seen as an attempt to transform implicit definitions (by postulates and by abstraction) into nominal ones. However, Peano himself pointed our Pieri as the inventor of the method: 'The late professor Pieri, and later Russell and Burali-Forti, already asserted that every definition by abstraction may be transformed into a nominal definition' (*1912a*, 403; also in *1915b*, 413).

We find the example mentioned by Peano in Pieri *1899a*. There the class of all the terms related is replaced by a common property based on an equivalence relation. To sum up, he introduced the *qualità* of the addition of two segments as the property belonging to any segment congruent to a given segment, and then he added this is equivalent to say: 'the "*addition* of |ab| and |bc|" is the "class of all segments congruent to |a'c'|"' (p. 216). Of course, this is a correct instance, but it contains no justification to prefer the suggested alternative and no theory about the new method. Fortunately, I discovered another example in a previous essay (Pieri *1898a*) containing a positive appreciation of the transformation into nominal definitions. It appears in a section about *order* where we find a theory and a notation on relations (see 3.5.1 above), after offering a definition by abstraction of the ordering direction (written $\sigma_{a,b,c}\ r$) as the common property to all coinciding orderings ('coinciding' was an equivalence relation previously defined). This is applied later to the sense (*senso*) of a straight line r. Then Pieri added (*1898a*, 37):

> this kind of indirect definition (definition *by abstraction*), in spite of being rigorous, is not preferable to a true and proper 'nominal definition', as for instance this one: 'Given the different points a, b, c, on the straight projective line, we call "sense (or *direction*) a, b, c" —written "$S(a, b, c)$"— to the "class of all natural orderings [*ordinamenti*] of r coinciding with $\sigma_{a,b,c}\ r$"'.

If we compare this example with the one mentioned above we find a clear preference for nominal definitions, the only 'true' one according to Pieri. Curiously, this passage was emphasized (among others) by Russell with a vertical line on the offprint he possessed since 1898 (see 3.5.1 above), which presumably constitutes a sign of Russell's knowledge of the method before the contact with Peano took place in 1900 (although of course we do not *know* the occasion when he actually read the article).

I come now to Burali-Forti. His writings from 1896 to 1900 show the evolution of his ideas about definitions and his increasing efforts to pursue a certain logicist idea. In *1896a* we can see clearly Burali-Forti trying to draw logicist consequences from Cantor *1895a*. Thus, after explaining Cantor's definition of cardinal number as an abstract entity common to all the equivalent classes, he stated something really like a logicist thesis: 'all properties of cardinal numbers are to be transformed into properties of *classes* and *correspondences*' (*1896a*, 35). After that, he wrote (in a clearer way than Cantor himself) about the way that 'the concept of number (finite or not) can be reduced to only the concepts of class and correspondence' (*1896a*, 37). However, in spite of offering logicist symbolical definitions (arithmetical addition in terms of logical addition; the number one through the unit class; the class of numbers in terms of the numbers of elements of all finite classes; see p. 51) he continued to present the number of a class as an 'abstract entity' (p. 39) and therefore depending on a definition by abstraction (p. 51; $\sim$ denoting one-one correspondence between classes):

$$N'u = N'v \ . = . \ u \sim v.$$

However, Burali-Forti was completely aware he was designing a new alternative to axiomatic arithmetic by Peano and Dedekind; for he alluded to Peano's arithmetic as an 'inverse treatment to ours' (p. 40).

1897a, the essay stating the famous (according to Russell) paradox, shows him maintaining definitions by abstraction, introducing the concept of *order type* as an 'abstract entity' that a class has in common with all ordered classes equivalent to itself (p. 159). In *1899a* we find the first attempt to define the three 'indefinables' from Peano (N_0, 0, successor) in terms of the notions of magnitude (*grandeur*) and operation (pp. 155-6), previously defined in logical terms. This anticipation led to *1899b*, where a new method arose concerning the introduction of derivative ideas in science.

The new way led him to an explicit attempt to avoid definitions by abstraction; probably suggested by Pieri's ideas from 1898 and 1899 examined above. The general method, applied to rationals, irrationals and cardinals, starts by defining equality in terms of membership to the same classes. This makes it possible to state several theorems with the common purpose of demonstrating that the necessary and sufficient condition to obtain the equality between two correspondences f and h (between two classes u and v) is that for every element x from u we have $fx = hx$ (*1899b*, 252). For cardinals it leads to a nominal definition allowing him to distinguish between *cardinal number*, a class, and *cardinal number of*, a correspondence between classes and single elements of the class *cardinal number*.

With that it is possible to overcome Cantor's view, limited to stating the equality of cardinal numbers in terms of the equivalence of the corresponding classes, and introducing the explicit concept *cardinal number of* by means of a nominal definition: 'one of the correspondences f between classes and single elements such that, for every class u, the classes v such that $fv = fu$ are all (and only) the classes similar to u' (p. 258). In this way Burali-Forti introduced a new notion in terms of other notions being (i) already known and (ii) *of logical* kind, so avoiding the abstraction process (in spite of maintaining the ground of abstraction: the equality $fv = fu$). The conclusion would be clearly logicist, but Burali-Forti limited it by maintaining the need for introducing primitive ideas by means of axioms.

The paper submitted to the 1900 Paris Congress was the culmination of this trend. There Burali-Forti held the view that nominal definitions are preferable to other kinds, arguing that they give *concepts*, whereas other definitions give only *intuitions* (*1900a*, 296). In general, he added, we use definitions by abstraction (or by postulates) when we do not know how to use nominal definitions. However the transformation, we are told, is easy. Concerning definitions by abstraction, Burali-Forti insists in the method already given in *1899b*. Thus, he defines the *direction of* as an operation *f* for straight lines such that, for every straight line x, the straight lines y such that $fy = fx$ are all (and only) the parallel lines to x (p. 296). However he also adds a nominal definition of the class *direction* itself in terms similar to those by Frege: 'the x such that there is a straight line such that $x = direction$ of a' (p. 296). However, the most interesting point appears when the author gives *nominal logicist definitions* of the three 'indefinable' terms in Peano's arithmetical system (N_0, 0, successor; pp. 298-9), now without resorting to the notion of magnitude, as in *1899a*. I have no room to quote them here, but it is important to say they are all constructed in logical and previously explained terms such as element, operation, class and similarity. The main conclusion is drawn precisely: it is no longer necessary to hold, like Peano, that number is abstracted from the five celebrated axioms. The nominal definition of the supposed indefinable terms shows them rigorously avoiding the abstraction process 'whose logical laws are not yet stated' (p. 301). The conclusion is that, although both definitions by abstraction and by postulates can lead to the usual notion of number, they achieve it in such a way that the 'nature and exact kind of N_0 are not stated' (p. 303). We can see then the exact parallelism with Russell's objections to definitions by abstraction (see 4.3 below).[1]

It is true that all this does not yet contain *exactly* the definition of the number of a class in terms of a class of classes similar to it. The only thing lacking was to join Pieri's method and Burali-Forti's efforts. The task was carried out by Peano himself in his *1901a* (see 3.2.4 above), although probably only to reject the break with the usual properties of cardinal number that the new definition presupposed (according to him). In the next chapter I shall explain Russell's reaction to all this, trying to establish the historical priority of his definition of number by considering the unpublished manuscripts, the global inheritance from Peano and the acceptance of Cantor's transfinites, by reconstructing the general context that made it possible to frame all these ideas into the philosophy received from Moore.

---

[1] Vuillemin (*1968a*, 178 ff) mentions the priority of Burali-Forti (with regard to Russell) in the *criticisms* against definitions by abstraction. Although such criticisms are important, the essential thing is to realize that Burali-Forti completely anticipated the logicist idea: to define in a logical way the supposed *primitive* ideas of Peano, which stated a 'gap' between arithmetic and logic. The reason seems to be that Vuillemin is convinced of the originality of Russell's logicist definition of number (see 4.3 below for a series of detailed arguments against this view). Torretti (*1978a*, 226) has pointed out, to my knowledge for the first time, the *independence* of Burali-Forti's logicist definition of number *in relation to Frege*; but unfortunately he does not consider the importance of this fact with regard to Russell.

4. The principles of mathematics

In this chapter I shall study the main *results* achieved by Russell once he was able to use the techniques of Peano's logic; then I shall compare these results with those from former unpublished manuscripts and publications. The relevant works now will be the writings, published or unpublished, from 1900 to 1903 and very especially *Principles*. However, I shall also resort to *Principia mathematica*, mainly to verify the lasting nature of these results, given that this work can be considered as the definitive version of Russell's 'mathematical philosophy'. At this stage the method is already clearly illustrated by an impressive list of constructive definitions, which show the great weight of Cantor's and Peano's techniques, which constituted a new viewpoint concerning definitions, as they were regarded according to Moore's philosophy, although the detailed study of these definitions, as regarded as philosophically 'constructive' ones, is left for the next chapter (see 5.3).

The following sections are ordered in the same way as the main parts of Peano's *Formulaire*, i.e. logic, arithmetic, analysis and geometry. Thus we shall be able to emphasize in the best way Russell's attempt to eliminate the three 'gaps' produced by Peano when introducing the new indefinables: cardinal number, real number and vector. I also include a first section about Russell's reaction to the Paris Congress of 1900 and a last one with the aim of establishing the Peanian heritage.

4.1. The reaction to the Congress of 1900

4.1.1. The notes to the manuscripts: from Moore to Peano

The history of Russell's reaction to the contact with Peano has been explained in a celebrated way by Russell himself in his autobiographical writings. However, in these accounts we often find the pre-eminence of a somewhat anecdotal viewpoint, with which it is difficult to reconstruct the link between the former and later periods to this event. In the following I shall offer several pieces of evidence in order to overcome this limitation.

On the day after his return from Paris Russell wrote to Moore: 'The Congress was admirable, and there was much first-rate discussion of mathematical philosophy. I am persuaded that Peano and his school are the best people of the present time in that line' (August 8, 1900; in O'Briant *1979a*, 183). Some months later, once he had a good grasp of Peano's symbolism, Russell described it in his letters to Couturat as 'excellent', as the cause of the achievements of the Italian school, and as an instrument of analysis capable to make the logical problems easy by avoiding paralogisms and simplifying formulas and demonstrations (letter of January 17, 1901). Likewise, Russell claims it is a true method capable of producing many new results (letter of February 1, 1901; both passages in Schmid *1983a*, 89), and this was the task he assumed for himself.

In the quoted letter to Moore there is an outstanding passage that emphasizes the importance of the relation among the concepts of variable, quantification, infinity and membership, in the framework of the former fruitless attempts:

> Have you ever considered the meaning of *any*? I find it to be the fundamental problem for mathematical philosophy. E.g. 'Any number is less by one than another number'. Here any number cannot be a new concept, distinct from the particular numbers, for only those fulfil the above proposition. But can any number be an infinite disjunction? And if so, what is the ground of the proposition? The problem is the general one as to what is meant by any member of the defined class. I have tried many theories without success.

The underlying distinction is, of course, that existing between inclusion and membership: the first one shows classes as indistinguishable from individuals, whereas the second can clearly distinguish them (in spite of containing the seed of dangerous paradoxes).

This was the most important technical achievement inherited from Peano. On the one hand, it made it possible, together with the universal quantification (also from Peano), to embrace the infinite, which led Russell to the acceptance of Cantor's transfinites (see 4.1.3 below); on the other, it led him to a more precise logic which avoided the confusion between class and individual, and made a more subtle analysis possible by providing instruments to define not only concepts, but terms, again through a Peanian notation: that for 'the'. (Grattan-Guinness discovered, in unpublished notes by Russell on Wiener's thesis of 1913, the only precise recognition we have that it was this particular point which impressed him *first* in the Congress; Russell wrote: 'It was a discussion on this very point [the need of a notation for "the"] between Schröder and Peano in 1900 at Paris that first led me to think Peano superior'; Grattan-Guinness *1975a*, 110.) Russell recognized, in general, all that (see ch. 6 of MPD), but I think he was completely unfair with his former unpublished attempts (see 2.1 above).

The logical calculus handled by Russell in the manuscripts was based on the whole/part distinction (i.e. inclusion), which had the merit of coinciding with the requirements of the Moorean analysis according to which the complex concepts are sets of constituent parts. However the logic of inclusion was at bottom syllogistic and it was incapable of solving the formidable problems of the foundation of mathematics, especially because it did not provide the mentioned distinction between terms an concepts (completely achieved by Peano). In MPD (ch. 6) Russell explains how before the Congress it seemed to him that Boole's logic was useless for solving the problems of the true 'grammar' of the arithmetic, and in a later letter to Jourdain (April 15, 1910; in Grattan-Guinness *1977a*, 133) we read:

> Until I got hold of Peano, it had never struck me that Symbolic Logic would be any use for the Principles of mathematics, because I knew the Boolean stuff and found useless. It was Peano's ∈, together with the discovery that relations could be fitted into his system, that led me to adopt Symbolic Logic.

However, only the private notes added by Russell to his last unpublished attempt (POM1) allow us to understand the transcendence of Peano's impact. I shall quote two of them which go much beyond the framework of the technical details. The first one is really crucial. It was added in the margin preceding the materials of the part II (devoted to 'Whole and part') in October 1900:

> I have been wrong in regarding the Logical Calculus as having specially to do with whole and part. *Whole* is distinct from *Class*, and occurs nowhere in the Logical Calculus, which depends on three notions: (1) implication (2) ∈ (3) negation. Whole and part requires the *Teoria della Grandezze*, i.e. a special form of addition, not that of the Logical Calculus. I must preface Arithmetic, as Peano does, by the true Logical Calculus, to be called *Book I*, The Individual.

We can already find here the basic elements of the adaptation of the preceding work to Peano's methods. First of all, we see that this adaptation did not suppose a complete rejection of the former work, but rather an attempt to take advantage of most of the material (especially of the last draft: POM1), even including the main plan and the order of the subjects. The main change was the introduction of *new indefinables* to replace the concept of number by a 'true' logical calculus capable of constituting a new viewpoint for the insuperable problems characterizing the logic used up until then. As we saw (in 2.2) that 'ontological' logic intended a direct handling of the terms by applying certain operations based on addition to them, and it was characterized by an analysis of the language which lacked the necessary conventionalism provided by Peano, so that it got lost in a mass of subtle distinctions and classifications in search of the authentic indefinables in themselves. The task made by Russell was to precede all that by a whole 'book' devoted literally to the Peanian logic, which led the analysis until the division of the concept itself of number (supposedly indefinable) into simpler elements. The proposed title of the new book (*The individual*) was finally rejected, but it is useful to imagine the sort of atomistic 'logicism' emerging from the former stage where it was latent.

The logical definition of number immediately provided the elimination of the first gap in the *Formulaire*, so that it was Peano himself who suggested the possibility of *reducing* the number of indefinables needed until then and, at the same time, the construction of *a unique chain of definitions*. Russell never admitted, however, that Peano's work involved *in itself* that possibility. The passage closer to such a recognition is —to my knowledge— this one: 'I saw that methods analogous to his would clarify the logic of relations, and I was led to the definitions of cardinal, ordinal, rational, and real numbers which are given in *The Principles of Mathematics*' (*1948a*, 137). In 4.3 we shall verify all that is underlying this obscure 'I was led'.

The second note added to the last unpublished manuscript (in the second folio of the mentioned chapter) shows all the philosophical importance of the distinction between inclusion and membership with regard to the deep *meaning* of such previous attempts:

> Peano's distinction of ⊃ and ∈ shows whole and part to be different from implication. The former is primary implication, the latter gives relation of simple part to whole. ⊃ between classes is derivative: its first significance is between propositions.

In this way the non-distinction (coming from Moore) between implication and presupposition (see 2.4 above) was definitively clarified with one stroke of the pen. Starting from this point, the first one could only appear between propositions and, although presupposition can play a similar role (e.g. between axioms), it can only take place *between concepts* (whereas a concept could imply another in the manuscripts). On the other hand, inclusion remained restricted to classes, so that, as we saw before, only membership will point out the distinction between class and individual. Finally, the converse of membership (the class abstractor or 'such that') will permit us to pass from classes to propositions (and vice versa; see 3.1.4 above).

As a whole, all this will permit us to pass from the rough Moorean logic to the sophistications of Peano and Cantor, although always within the same framework: definition (and analysis) as the genuine method and external relations (now explicit) as the general context.

4.1.2. The first writings

From 1900 to 1903 Russell published several articles of unequal interest and wrote others that were never published. Here I shall refer only to the general trends —which will be studied in detail in the following sections— leading to the culmination of POM and PM.

Relations were somewhat already fundamental in the unpublished writings previous to the Congress of 1900. In fact Russell had even arrived at a first attempt of classification according to the several properties exhibited by them (see 2.6 above). They constituted the best recourse for pluralism and, with that, for the incipient logical atomism. The novelty now will be to transform relations into *elements* capable of belonging to certain classes, i.e. to make the construction possible of a true *arithmetic* powerful enough, through the corresponding *operations*, to give an account of concepts as important as Cantor's correspondences (mappings) (which gave rise to the definition of cardinals) and the notion of order (which permitted him to give an account of series, progressions, ordinals and mathematical induction). Thus, relations were the natural (and to some extent previous) framework where the alliance took place between the mathematical logic from Peano, the valuable results of Cantor, the ordinal theory of Dedekind and Peano, and the new method of mathematical definitions, which were destined to absorb the old 'philosophical' definitions based on intuition and ordinary language (as it was practised, e.g. in PL; see my *1990k*).

Russell *1901a*, the first paper published in Peano's journal, marks the first assimilation of the new methods. However although it offers a new treatment of relations and the seed of the principle of abstraction (as it was formulated later), it does not yet contain a satisfactory logicist definition of the cardinal number (in fact the only —verbal— logicist definition was added in proofs), nor a systematic transformation of definitions by abstraction into nominal definitions. To understand the origins of these two recourses we shall have to resort to the unpublished versions of this paper (see 4.3.1 below).

On the other hand, *1901d*[1] was a true manifesto leading the logicist idea much beyond Cantor, Peano or even Frege. There are three basic points. The first one, of course, is the thesis according to which mathematics can be reduced to logic, although the particular formulation is similar to that of B. Peirce and Whitehead (see 1.1 and 1.3 above): 'Pure mathematics consists entirely of assertions to the effect that, if such and such a proposition is true of *anything*, then such and such another proposition is true of that thing' (*1901d*, 75-6). Therefore mathematics is logic from its foundations themselves, for the latter provides the rules of inference to be applied in the former to draw 'necessary consequences'. This definition, that was repeated —in the essentials— in POM, gave rise to regard mathematics as 'the subject in which we never know what we are talking about, nor whether what we are saying is true' (*1901d*, 76). In any case the logicist thesis is stated clearly: 'All pure mathematics —Arithmetic, Analysis and Geometry— is built up by combinations of the primitive ideas of logic, and its propositions are deduced from the general axioms of logic, such as the syllogism and the other rules of inference' (*ibidem*). Like Burali-Forti (see 3.5 above), Russell insists that this reduction has to be made by defining the three arithmetical ideas supposedly primitive by Peano (*number*, *zero* and *successor*) in logical terms, i.e. in terms of 'classes and relations', although a new emphasis in relations is added (*1901d*, 79).

The second basic point consists in the importance given to the *elimination* with regard to logical definitions. Russell approves of the destruction of the notion of infinitesimal made by Weierstrass (and confirmed by Cantor and Peano), under an Ockhamian argument: 'everything could be accomplished without it' (*1901d*, 83). Thus, since Weierstrass built the concept of limit in a way independent of the customary spatial intuition, he set up the basis for a logically rigorous mathematics. On the same line, Russell praises (in very general terms) the definitions of continuity and infinity by Dedekind and Cantor under the argument that they introduced a great rigour and precision into such notions (*1901d*, 85, 90).

The third important point is the logicist reduction of geometry, to which Russell alludes here for the first time (although in FG he had started a line of work partially similar; see 1.5 above). Now he tries to overcome his previous attempts (in particular his *1899b*; see my *1990a*) by holding the thesis that geometry depends upon order. In particular it can be defined as the drawing

---

[1] Written in January 1901, according to the letter of 1910 to Jourdain already mentioned (in Grattan-Guinness *1977a*, 133).

of consequences from certain axioms about series of more than one dimension (*1901d*, 92). According to Russell, this logical reduction, rather than non-Euclidean geometries, will ruin Kant's philosophy (*1901d*, 95). In this way, we can regard this little article as a true plan of the later basic structure of POM.

There is, however, another relevant article as regards the reaction we are explaining: *1901c*, which can be regarded as the updated version of the paper actually submitted to the Congress (*1900b*). The comparison between both versions shows the new influences. As I explained elsewhere (my *1990k*), *1900b* was an attack against the relational theory of space and time, that was then regarded as too inclined to monism and idealism. The changes affect only the two first parts (space and time), and the rest remains untouched,[1] but it was sufficient to modify the general present goal: to show, not only that the relational theory is wrong, but also that points and instants are not mathematical fictions. Moreover, the new constructive trend is emphasized on going beyond the purely 'logical' framework by offering an *analysis* of the involved concepts and 'constructing a logically permissible account' that can destroy the belief according to which space and time are mere appearances (*1901c*, 294).

As for time, the reduction of order to relations vanishes (although it was taken up again in *1901b*; see 4.3.4 below); the absolute theory continues to be defended, but now by insisting that symmetrical and transitive relations are *analyzable* (Russell adds he had 'demonstrated' this in *1901a*). Starting from that, the conclusions to *1900a* (which were rather critical) are modified by offering an implicit application of the principle of abstraction (which we shall consider in detail in 4.3.4 below). Concerning space, now the constructive viewpoint has the pre-eminence over the critical one. The ground is the definition of space as the class of entities (points) whose relations are described using Peano's axioms (from *1894a*) as a guide. In this way Russell arrives at three indefinables: the point, the class of relations defining the straight line, and what is called 'magnitude of divisibility', although only the latter is presented, somewhat strangely, as a general logical notion (with the argument that it is referred in no way to space). On the other hand, the straight line is not claimed as being reduced to logical notions, despite that it is defined through relations (POM would take this latter step).

The important *general trait* in all that is the form in which the Moorean logic was used before the Congress, i.e. rather as a critical instrument destined to discover and analyze distinctions often confused by traditional logic, but with no constructive technique going beyond a view of relations as genuine terms or logical atoms. However, once Peano's methods led Russell to the possibility of considering relations as logical constants, the constructive and technical task had to be undertaken. Since the general framework was Moore's 'logical atomism', the corresponding absolute ontology was constituted as a difficulty to be overcome rather than to be used as a help to work. Russell had to hold this ontology, which was presented as the only valid alternative to his realism and anti-idealism, whereas, on the other hand, he was led through the logicism inspired by

---

[1] This third part became chapter 51 of POM, which has not been always realized by some commentators who regard this material as being new despite the fact that it proceeds from the stage previous to the Congress.

Peano to the construction of *all* geometry; i.e. he must try to preserve the points as irreducible entities and, at the same time, he must try to reduce them to logical entities. Something similar took place with relations: for Moore they had to be emphasized as ultimate material; for Peano they had to be defined as classes of pairs. As for points the pre-eminence was for logicism (although with the difficulties we shall see), but concerning relations the pre-eminence was for Moore's viewpoint (although later Russell had to change in favour of Peano; see 4.2.2 below).

4.1.3. The acceptance of Cantor

In 2.8 I tried to explain why Russell rejected Cantor's transfinites; now I shall try to explain why he accepted them. However, this has to be made not simply by presenting this acceptance as a part of Peano's logic (see Grattan-Guinness *1980c*), but as a consequence of a more general change affecting the whole context.

Firstly, the reasons that made the acceptance impossible begin to disappear. The demand of an immediate intuition of the indefinable terms (as appeared in PL) is partially modified by the acceptance of certain interdefinability of the logical 'simples'. With that the constructions became easier and the intuition seemed to remain relegated to 'philosophical definitions', although the problem subsisted that logic had to be the ground of both philosophy and arithmetic. In fact, Russell seems to be still dubious about constructions, perhaps because of the old Bradleian distrust of them; that is why he distinguishes between definitions and constructions, assimilating Peano to the former and Cantor to the latter. Moreover, the conventional element in Peano's logic must have made the rare Cantorian mixture of Platonism and conventionalism easier ('mathematics is free'), which, again, appeared as something more admissible when the demand of an immediate intuition disappears. On the other hand, the 'leap' toward the second number-class was already unnecessary starting from the definitive version of the Cantorian construction (*1895a*), where through the pre-eminence of cardinals, the mechanism of the correspondences (mappings) and the logical addition, one arrives at the concept of order type and almost at a logical definition of number (see 1.2 above).

Secondly, there were also positive reasons, despite the fact that Russell himself tried to minimize them when considering the role of Cantor in his admission of the new logic: 'The only part played by Cantor's work was that I tested my logic of relations by its applicability to Cantor' (in Grattan-Guinness *1977a*, 134). It is obvious that the discovery of the gap about Cantor in the *Formulaire* must have constituted a true challenge for Russell (as I showed in 3.1); in fact two of the first important Russell's articles (*1901a* and *1902b*) were partially devoted to these matters. However the *content* itself of Cantor's work also thoroughly impregnated Russell's logicism; and that because, among others, the thesis that mathematics is reducible to logic included Cantor's set theory (Grattan-Guinness *1977a*, 157). From this point of view it seems impossible that neither POM nor PM can hold without Cantor. That is why I pointed out before that we must speak of a general framework explaining, at the same time, the assimilation of Peano and that of Cantor.

From the viewpoint of the manuscripts prior to the Paris Congress it can be said that the true revolution was the acceptance of Cantor. After all Peano made it possible to deepen the power of definitions, which was the ideal method for Russell from 1898 on; but Cantor required, not only a 'logical' definition of number, but also a general conception that it made the distinction finite-infinite unnecessary from the beginning, which, on the other hand, provided to the system with a generality almost irresistible for a mathematician. Already in 2.8 I mentioned a passage of POM1 where Russell rejected Cantor, whose beginning was preserved in POM, but replacing the rejection by an acceptance. Let us now consider it with some more detail.

Chapter V/5 of POM1 begins by recognizing Cantor's merit in having created the mathematical theory of the infinite as something independent from the infinitesimal calculus; and in having bravely faced the problem of the infinite by pulling the skeleton out of the cupboard. Up to this point the passage exactly coincided with the beginning of §283 in POM, but later on the two versions differ. POM1 continues by showing his disagreement, in spite of recognizing the new theory as something prior to the Calculus, to finish by denying that it solves none of the philosophical difficulties of infinity (in 2.8 I quoted the passage). However, the version of POM adds (§283):

> Indeed, like many other skeletons, it was wholly dependent on its cupboard, and vanished in the light of day. Speaking without metaphor, Cantor has established a new branch of Mathematics, in which, by mere correctness of deduction, it is shown that the supposed contradictions of infinity all depend upon extending, to the infinite, results which, while they can be proved concerning finite numbers, are in no sense necessarily true of *all* numbers.

In this way Russell accepted the old argument by Cantor (and Couturat) and the correctness of the demonstrations where he saw before only skillful tricks. Therefore, the kernel of the acceptance lies in the logical link between the construction of infinity, symbolic logic and logicism, which are mutually supported.

In fact, already in *m1900b*[1] this link was clearly stated. Thus, when Russell (in a never repeated way) admits that the logicist viewpoint for arithmetic can be found in Peano as a possible alternative, he adds that Cantor's transfinites proceeded from this 'other' point of view according to which numbers are to be based on definitions constructed out of classes and series (*m1900b*, 13). In this way he immediately realizes that the natural point of departure of Cantor's construction is logic, with which he is thinking of the relation of correspondence between sets or classes. Therefore the attempt to obtain a unique chain of definitions involves the reduction to a unique set of indefinables that, at the same time, can serve to define number, no matter whether finite or infinite. A consequence of this procedure was, to some extent, a spacing out from Moore in so far as the technical constructions obtained the pre-eminence, but after all logic remained as the only

---

[1] Kenneth Blackwell has confirmed my opinion and has said to me that the composition of this manuscript proceeds from October 1900.

ground, despite that logic was then very different from that of Moore. Therefore, logicism, Peano's logic and transfinites constituted *a unique basis* for all mathematics.

Russell felt so sure about Cantor's construction that he wrote: 'it can be proved, from the general principles of logic, to be the only possible theory' (*1903c*, 268). Starting from this conviction it can be explained that in the first two articles published after the Congress of 1900 (see 4.1.2 above) the acceptance was already indisputable. On the one hand, it made it possible to see the three fundamental problems (infinity, infinitesimals, continuity) as *one and the same problem* capable to be reduced to the concept of order (POM, §179). On the other, it permitted Russell to define arithmetical addition of cardinals (once they had been constructed as classes of classes) by means of logical addition, in a way that the arithmetical sum of two cardinal numbers is the cardinal number of the logical sum of the corresponding classes (which takes place already in *1901a*).

Another advantage of such a viewpoint was that, as the involved addition was a *logical* one, it was prior to order (and quantity), so that it stated the pre-eminence of cardinals. It is true that Russell showed some doubt on this primacy (see 4.3.5 below), but the general line of the manuscripts presupposed it, starting from Couturat and Cantor himself. In this way one can reduce ordinals to merely finite numbers (in being the result of *successive* additions), so that mathematical induction remained relegated to ordinal numbers, as Burali-Forti had claimed (in his *1900a*, 301). That is why his logicist construction, in being based on cardinals (no matter whether finite or infinite), was identified by Russell with Cantor's (POM, §109), with which he was stating that the theory of infinite numbers is *simpler* than that of finite numbers because it does not require mathematical induction.[1]

Moreover, Russell rejected the definition proceeding from Bolzano, Dedekind and Cantor according to which the infinite must be regarded as *ordinal*, probably thinking of Cantor *1883a* where ordinals appear before cardinals because of the 'first principle of generation' (despite the epistemological priority of cardinals). For this reason he made the effort to offer an alternative definition lacking this 'defect', although the final definition was lexical rather than constructive (see 2.5 and 2.8 above). It is easy to imagine Russell's surprise in reading Cantor *1895a*, where the problem was overcome, and in knowing the various Peanian alternatives to define infinity starting from order. As we shall see (in 4.3.2) Russell took advantage immediately of the expressive possibilities of Peano's symbolism concerning this subject. Already in 1900 he wrote, for example, that the notion of membership makes it possible to deal with infinite classes by allowing the symbolical expression of *any*: it is sufficient with stating membership to an infinite class as the *hypothesis* of an implication, as for instance: $x \in N . \supset_x . ...$ (*m1900b*, 6). In POM

---

[1] As it was clearly seen by Couturat (*1904a*, II-III, 228). In one of the notes added after the Congress to POM1 (I/4) Russell already recognizes that the proofs by induction fail with transfinites, which cannot be obtained through successive additions. In POM (§109) he summarizes the new method with regard to the customary one into three points: (i) Cantor's generality, which allows the handling of numbers whether finite or infinite; (ii) the logical definition of arithmetical addition; (iii) the definition of the first ordinal. Thus, the perfect compromise between Cantor and Peano appears.

the link is confirmed by claiming that the main antinomy of mathematics (that of infinity) seems to be solved by means of a suitable philosophy of *any* (§179), and by showing the independence between the whole/part relation and enumeration, which become possible through a relation between implication and the variable (§342):

> If *a* be a class-concept, an individual of *a* is a term having to *a* that specific relation which we call the class-relation. If now *b* be another class such that, for all values of *x*, '*x* is an *a*' implies '*x* is a *b*', then the extension of *a* (i.e. the variable *x*) is said to be *part* of the extension of *b*. ... The definition of whole and part without enumeration is the key to the whole mystery. The above definition, which is due to Professor Peano, is that which is naturally and necessarily applied to infinite wholes.

However, the need for changing the logical structure of the manuscripts appears (see 4.1.1 above) as soon as the whole/part relation is so defined, since the kind of logical calculus needed has to be independent of inclusion, so that it must be logically preceded by the *intensional* viewpoint, which can only provide propositional functions, i.e. predicates. Thus, as in considering infinite wholes the class-concept is needed, then 'the theory of whole and part is less fundamental logically than that of predicates or class-concepts or propositional functions' (POM, §139). That is to say, the admission of infinite wholes makes the whole/part relation to lose its property of being simple (indefinable), so that it cannot be already located in the beginning of the construction of mathematics as a part of its foundation (as Russell believed in the manuscripts). Now propositional functions have to precede the authentic logical calculus and (as the basis of the logical definition of cardinals) even arithmetic. Thus we come to the link between infinite numbers (as simpler objects) and Peano's logic. Logicism, again, becomes the only general framework allowing Russell to lead Cantor and Peano to their ultimate consequences. Later on we shall see the rest of the methodological advantages.

4.2. Logic

4.2.1. The indefinables and the propositional function

As a whole we can say that Russell's logic at this stage (essentially in POM) constitutes a compromise between the philosophical ideas from the unpublished manuscripts and Peano's techniques, so that in adding to POM1 a new part, with the new logical indefinables, Russell tried to preserve the maximum possible number of ideas from that last unpublished attempt (which was almost completed at the moment of the Congress). In particular, he intended to preserve the reductionistic pattern inherited from Moore. This viewpoint gave rise to a main problem: to look for compatibility between (i) the relational theory of judgment and the vision of the indefinables as concepts (terms) in themselves, and (ii) the theory of judgment underlying Peano's logic, very close to the subject-predicate pattern, and a vision somewhat relativistic of the indefinables. The compromise affected in the same way the theory of classes and relations, especially concerning extensionality. As a result of all that, logic could be regarded as Peanian in the essentials but Moorean in the philosophical ground. I shall emphasize in the following only the points that more clearly exhibit this compromise, considering as presupposed that in all the rest Russell followed the inheritance from Peano (as he himself recognized in POM, except for relations), in spite of having introduced some new terminology: 'propositional function', 'formal implication', etc. As for the literature, only Vuillemin *1968a* has studied the logic of POM, although only from a technical viewpoint, whereas Wang *1967a* and Gödel *1944a* took into account almost only PM.

It is difficult, if not impossible, to offer a complete list of the indefinables in POM. Russell himself seemed to elude it by mentioning *several* lists according to different places in POM (§§1, 12, 106). In part because, as we shall see, he did not achieve a clear elucidation for the status of the indefinables (which had been regarded as to some extent arbitrary by Peano); in part because he did not clearly distinguish between logical constants, logically presupposed 'notions', and certain terms playing both linguistic and ontological roles (i.e. assertion, denotation, definition, etc.). This ambiguity is already a sign of the compromise mentioned above.

If, in spite of these obvious difficulties, we try to obtain a list of indefinables, we have to admit, at least, the following. In the calculus of propositions (§§14-19) we find *two*: material and formal implication, which constitute the ground to define equivalence (mutual implication), disjunction and negation. The logical product is defined in a complicated way starting from one of the ten axioms (or Pp) that rule this calculus, which includes six of the rules of inference used by Peano. Of course, these definitions already give rise (regarded as constructive definitions) to the problem of the distinction between what is technical and what is philosophical (see 5.3 below).

In the calculus of classes (§§20-28) *three* new indefinables appear: membership, class-abstractor ('such that') and propositional function (involved, as we shall see, in the formal implication). Two Pp are also added and, following Peano, three logical concepts are constructively defined: identity (as membership to the same classes), the existence of a class (as having at least a term), and the null-class (in several ways). Finally, in the calculus of relations only *one* more indefinable appears: the idea of relation, through six more Pp (and two complementary Pp attributing the idea of relation to material implication and membership).

To these *six Pi* Russell later adds three difficult ideas: *truth, constancy of form* and *denotation*. All these ideas give rise to many problems (likewise some of the former ones), especially when other notions appear which are not regarded as primitive, but it is not clear whether they can be 'constructed' within the system (e.g. proposition, assertion, class, variable and definition, among others). Curiously, most of such problems arise with regard to the indefinables introduced by Russell himself (as for instance with the ideas of denotation and relation), or in relation with the 'analysis' of some of the ideas proceeding from Peano (as formal implication). Before considering these problems we must try to elucidate the ontological status that Russell intended for the indefinables (and indemonstrables).

Once decided that the concepts of mathematics are to be reduced to logical fundamental concepts (POM, preface), Russell's viewpoint is apparently similar to Peano's, for he adds that such concepts are the particular notions appearing in logical propositions (§12). Thus, although they only can be given by *enumeration*, in practice they can be *discovered* through an analysis of what is really shown in such proposition (§10). However, as I explained in detail in 3.1, Peano was very soon convinced of the interdefinability of such constants, which led him to maintain the arbitrariness of this or that selection of them as *Pi*. However Russell could not follow Peano up to this point if he wanted to preserve the supreme feature of Moore's ontology. Besides, the technical possibilities of interdefinability seemed to allow a type of 'deeper' analysis where certain notions can be presented as *authentically* simple, which led Russell to speak about two kinds of logical constants (connectives): the definable and the indefinable ones (POM, §§12, 1). In this way Russell could accept Peano's thesis of arbitrariness *and* maintain true indefinables for every particular selection of them.

With this aim, Russell resorts to the following arguments. He firstly begins with the mathematical possibility of defining 'simple' terms by means of fixed relations with other terms, i.e. through the construction of a concept *denoting* certain term in an ambiguous way. Thus, the notion of denotation has to be taken into account in every selection, which seems to stop (to some extent) the arbitrariness (§31). Moreover, some practical reasons are added: the selection of a small number of indefinables is useful because of the precision and clarity; besides, it allows it to demonstrate that this set of constants is *sufficient* to draw certain consequences (§32). The slight clarity of Russell's position is brought to light when he finishes by adding that although any selection is arbitrary, however there must always be some indefinable logical constants (§108). With which one does not clearly understand whether he means that we should not know how to define them (although they would be definable in themselves), or that in fact some constants are necessarily indefinable if we have to start from a set of *Pi* and *Pp* powerful enough to give an account of mathematics. However, when he emphasizes that some particular constants (e.g. membership) are indispensable he seems to choose the second possibility, which is confirmed in writing: 'the indefinables of Pure Mathematics are all of this kind'.

Once the indefinables have been technically proposed, the second kind of analysis must begin. Through it, Russell tries to *explain* all the indefinables to 'show' them in order that the mind can recognize them as simples (like a colour or a taste). Only this second kind of analysis will grant the complete objectivity of such contents (to remember the term used by Bradley). Therefore the need for this analysis is pointed out in due course (POM, §3), although Russell does not tell us about the possibility of applying it to *definable* notions.

A problem seems to underlie this position: on the one hand the distinction between philosophical and mathematical definitions is accepted, whereas on the other the need for them is not mentioned in some places (especially in the technical part). This seems to be the proof that the compromise between Peano and Moore does not work; even so the philosophical definitions have to be maintained to grant that some entities are indefinable in themselves, as it is shown by the fact that they are presented as such to the intuition. We shall see later that Moore's theory of judgment was hardly compatible with Peano's logic and Russell's propositional functions.

I come now to the *practice* of this kind of philosophical analysis (or explanation), i.e. to show the problems arisen as soon as one tries to evaluate the definability of the notions declared as non-primitive. This is the most interesting part because it *shows* the deeper difficulties of the attempt. Here the idea of *propositional function* is the most important because (i) it is closely related to all the rest of them, (ii) it seems to involve the traditional theory of judgment based on the subject-predicate pattern (perhaps including Frege's analysis of the proposition into function and argument), which Russell had to face, (iii) it is presupposed by the ideas of formal implication and that of variable.

However, Russell defines propositional functions by means of the variable (see 3.1 for the Peanian distinction between apparent and real variables): 'where there are one or more real variables, and for all values of the variables the expression involved is a proposition, I shall call the expression a *propositional function*' (POM, §13). That is to say, φx is a propositional function if, for every x, φx is determined as a proposition. Nevertheless Russell adds that this is not a true definition, but only an 'explanation' (§22), probably because (as we shall see) it presupposes the idea of 'all', that is indefinable, and also that of proposition (or even that of assertion), that is doubtfully definable.

The link with the predicative pattern is immediate. Russell cannot accept Peano's interpretation in terms of membership because it seems to him that it would lead to the non-distinction between predicate, class-concept and class, and he needs to distinguish between these two last notions, for they are the ground of the ontological distinction between terms and concepts (see 5.4 below). That is why Russell wrote: '"x is a u" is a propositional function when, and only when, u is a class-concept'. This leads to distinguishing the subject from the predicate, or at least the subject from anything not being a subject. For this reason one needs only to replace, in any proposition without real variables, one of the terms not being a verb or an adjective by other term to obtain a propositional function (§22). This latter prohibition is already a strong (though implicit) argument against the relational theory of judgment because this theory did *not* permit the distinction between subject and predicate.

Russell tries to minimize this unavoidable consequence by proposing that propositional function involves the verb as well, whereas *class* would remain reduced only to the adjective or predicate (POM, §55). This allows him to analyze propositional function into a variable element (the 'place' of the subject) and a constant one (the assertion), which embraces all that not being part of the subject in the proposition. However, the attempt was discarded because it requires the addition of the notion of 'all' (which has to embrace the terms capable to replace the variable element) to the constant (or indefinable) element, which is declared as very complicated (§81), and also because of the relational propositions, whose 'sense' would vanish within such a pattern (§82).

It is really paradoxical that Russell falls into the predicative pattern, that he is trying to avoid, by resting on a notion (that of propositional function) that is after all indefinable and can be analyzed in every case except for relations, where Peano's procedure (the classes of pairs) would fail according to Russell. However the main reason underlying this argument is Russell's conviction that the Peanian expression '$x \in a$' must be framed into the predicative pattern, although Peano really made use of it from an extensional viewpoint in which predicates are to be transformed into classes without passing through class-concepts (see 3.1 above). Russell thinks that it is necessary to introduce propositional functions in the calculus of classes just because he interprets this notion as an alternative to the functional viewpoint of relations (POM, §22):

> Peano does not require it, owing to his assumption that the form 'x is an a' is general for one variable, and that extensions of the same form are available for any number of variables. But we must avoid this assumption, and must therefore introduce the notion of a propositional function.

However, the new notion is also incapable of giving rise to relations in Russell's system, which seems to be a proof of its dependence upon the predicative pattern. Russell recognized that propositional functions are more important than other functions and even than relations themselves (§254), but until his contact with Frege he did not accept the extensional pattern of Peano (and Peirce), as can be seen in the appendix to POM (§485) and of course in PM (where the notation 'xRy' was preserved only because of practical reasons).

In the rest of this section I shall briefly refer to the other notions related to propositional functions. First of all to the *constancy of form*: 'There is, for each propositional function, an indefinable relation between propositions and entities, which may be expressed by saying that all the propositions have the same form, but different entities enter into them' (POM, §33). Therefore every propositional function defines a class of propositions sharing the same form. Curiously, however, Russell does not apply here his well-known *principle of abstraction* to replace the relation 'to have the same form' (an equivalence relation) by the class itself; in fact he introduces this form as a *Pi* (§86). Perhaps he was afraid about some paradox. In any case he avoided the rejection of classes at this stage, although he recognized that the class of the propositions having the same form is more important that the notion of class in itself, because the latter can be defined through the former, but not vice versa (§106). With that he was thinking of the (also indefinable) notion 'such that': a propositional function gives rise to the class of propositions that it defines, and also to the class of the terms 'such that' satisfies it.

This leads to the notion of *truth*, that Russell introduces as an indefinable property of propositions by avoiding analyzing it (although he does not resort to Moore's intuition, which was only one more 'concept' or term; see my *1990f*). Instead of that, Russell limits himself to adduce that mathematics only *makes use* of this notion, which is not an element of its propositions (POM, §1). Perhaps he was afraid of the danger of psychologism if he identified the truth with the fact of the assertion itself (§38), with which it would be necessary to distinguish between the 'logical' assertion from the psychological one (§52) (in fact he was forced to make this distinction because he still conceded genuine reality to false propositions). It is also possible that here Moore's vision of the proposition as a complex concept underlies, which is opposite to the kind of proposition needed as a ground for propositional functions, which (as we saw) presuppose the subject-predicate distinction. Thus, to transform truth into one more concept it was required to regard the proposition as a concept composed of other concepts belonging to the same ontological level, which was impossible through the distinction between terms not being concepts (things) from those being so. Hence the need for admitting truth as a primitive idea, despite the fact that this endangers the independence of the notion of propositional function, which requires truth to give rise to classes in a valid way (see Griffin *1980a*, 125 f, for an interesting dilemma between material and formal implication).

I come now to the notion of *proposition*. Russell regards it as derivative (which supposes a certain progress with regard to Peano, for whom it was neither primitive nor derivative) but he offers two somewhat unclear definitions. The first one rests on the notion of truth: a proposition is anything capable of being true or false (POM, Preface, §13). With that we arrive at the problem mentioned above, for the only relevant truth for logic is formal truth, and only propositional functions (which give rise to formal implication) have the property of generating classes of 'true' propositions through certain forms. The second definition resorts to the notion of implication: 'Every proposition implies itself, and whatever is not a proposition implies nothing. Hence to say "p is a proposition" is equivalent to saying "p implies p"; and this equivalence may be used to define propositions' (§16). It is another example of a 'construction' that is limited to point out a necessary and sufficient condition; that is why it gives rise to some problems (the same problems avoided by Peano in eluding the definition).

A first problem: it makes the implication between concepts impossible, whereas it was usual in the manuscripts. Besides, to achieve this limitation the idea of truth is presupposed, in spite of the fact that (as we shall see below) Russell wanted to arrive at the idea of implication in an independent way from truth. Another problem is related to equivalence. Since it is defined as reciprocal implication, to say that an implication ('p implies p') is *equivalent* to a given proposition ('p is a proposition') supposes it to remain within the framework of implication. Finally, if we define proposition by means of implication, it seems difficult to understand the meaning of the latter in an independent way, especially by taking into account that Russell denied the possibility of defining it —like Peirce or Peano— in terms of truth.

To sum up: proposition, truth and implication can be intuitively understood only as a whole. It seems, therefore, that Russell's 'second' kind of analysis does not manage to *show* before the mind the simplicity of these notions in an independent way from their technical (mathematical) working. Perhaps this was the reason why Russell, on starting arithmetic, intended to have nothing to do with the ballast of the philosophical definitions. However, the fact that he undertook the long and detailed analysis we can read in the first part of POM shows that he felt forced to apply Moore's analysis to the basic indefinables of the system.

I come now to the fundamental notion of *implication*. I shall begin with material implication, referred to above with regard to equivalence and truth. By following the line stated in the manuscripts prior to the Congress of 1900, Russell insists that implication is independent of truth according to three reasons: (i) if '*p* implies *q*' then 'the truth of *p* implies the truth of *q*', hence we obtain new implications but not the definition of implication (POM, §16). It seems, however, to be only a verbal difficulty: we need only regard the two referred propositions as equivalent, or introduce the propositions stating the implication between the falsehood of *q* and that of *p* (this was the procedure followed in *1906a*, 162); (ii) the equivalence cannot be admissible for it consists in a mutual implication, hence Russell rejected the definition based upon the proposition '*q* or not *p*' in spite of recognizing the 'structural' equivalence (§16); (iii) 'disjunction... is definable in terms of implication' (*ibidem*).

All that seems to be a consequence of the belief that the interdefinability makes the notions invalid and that (as we saw before) there must be some constants being indefinable in any logical system. In PM the conventionality prevailed, in admitting the standard definition by means of negation and disjunction (*1.01). Thus, only in PM was it possible to replace the equivalence by the sign of definition (=Df), starting from a viewpoint regarding the definition as a mere technical recourse affecting only the *symbolism* (in this way Russell tried to go beyond Peano himself, for whom definition was a *Pi*). Again we verify that the influence from Moore constituted a ballast to the setting up of a true logical system.

Finally, I shall briefly refer to *formal implication* in relation to the two remaining notions: denotation and the variable (I included a detailed study of Russell's concept of denotation and its evolution in my *1989b* and *1990h*). First of all, this notion starts from the same idea that material implication; the only difference is that the latter takes place between propositions whereas the former between propositional functions. But, as a propositional function does not imply anything by itself, the theory has to be complemented with the analysis of the variable, which, again, led to the problem of denotation. That is why Russell prefers to regard the expression '$\varphi x . \supset_x . \psi x$' as a unique propositional function and not as a relation between propositional functions, with the argument that if we replace the variable for 'any' term, we always obtain a true implication (POM, §42).

However, this unavoidably led to the incorporation of all the problems mentioned above concerning the ideas of proposition and truth, and to relate formal truth to quantification. Besides, if any term can replace *x*, there must be a *constancy of form* founding the implication itself, and this seems to be *assertion*, i.e. what remains in a proposition when one dispenses with the subject. The result is that formal implication takes place between assertions (POM, §44); but assertions involve only *concepts* (and not terms of subjects), so that one needs to admit that which was before denied: there can be implication between concepts. This leads directly to a logic based upon inclusion and the corresponding problems.

For the same reason, the fact that one can replace *x* for any term gives rise to a whole class of implications, i.e. to a 'variable implication' (POM, §42). In any case, the concept of variable is brought to light as something hardly distinguishable from the structure generated by any concept ambiguously denoting (*all*, *every*, *any*, etc.). However Russell's analysis of the variable identifies it with the object denoted by 'any term', which gives rise only to an ambiguous individuality, which is always the result of a propositional function. Therefore formal implication rests on the ideas of variable, denotation and propositional function (§93), whereas the variable itself presupposes the two other (§86) and, again, propositional function cannot be explained without a reference to the rest of them. The second kind of analysis leads us, again, to the same situation. As we shall see below it seems that the logical techniques were hardly compatible with Moore's linguistic and ontological analyses, so that the 'guidance' Russell intended to discover in grammar (§46) gives rise to problems rather than to solutions.

In *1906a* the problem was simplified in admitting the following as indefinables: variable, assertion, implication and propositional function, which doubtlessly supposed a closer approach to Frege. In PM propositional functions generate both classes and relations, which result immunized against paradoxes in being declared as 'incomplete symbols', together with descriptions (see my *1989b*). Moreover, by introducing the proposition as a *Pi* (PM, *1) and defining implication in terms of negation and disjunction, truth is transformed into a harmless idea, as it is customary in a more extensional logic. Finally, universal and existential quantification are rigorously stated (*9) (also here Frege has the pre-eminence over Peano) and constitute a new basis for the variable and formal implication in being demonstrated that, through its only application, that which is valid for elementary propositions and their material implications can be extended to more complex propositions (with which the problem of denotation disappears, once the dubious entities are eliminated through the theory of descriptions). However, all these novelties were already completely foreign to the old logic inspired in Moore. We shall see that the same thing happened with relations.

4.2.2. Relations

Already in the unpublished manuscripts, relations were regarded as external, irreducible and needed for overcoming the contradictions proceeding from monism (see 2.6 above). The relational (non-existential) theory of judgment forced Russell to present concepts as realities independent of existence (see 1.4 above and my *1990k*). Russell's efforts to achieve a classification of relations (e.g. *m1899a*) still did not allow the construction of a calculus, in part because of the lack of a convenient notation, in part due to Bradley's objection against the relation between the relations and the related terms.

The contact with Schröder and Peano provided Russell with a method to make the progress in this field possible; however, his own philosophical project prevented him from the practical and theoretical acceptance of the reduction of relations to the functional pattern and the ordered pair. According to Peano it became possible to handle relations through membership and implication, which allowed it to regard them as individuals and classes from which to state certain properties.

Besides, the class abstractor permitted Russell to make use of the propositional function pattern to determine such properties, as for instance those referring to the related terms:

$$r \in rel \,.\, \supset \,.\, \rho = \hat{x}\,(x\,R\,y),$$

which could be already taken as a point of departure for a calculus.

However Russell needed to interpret relations as ultimate realities being completely independent from the related terms, so that he could not accept them as properties of those terms (e.g. $R(x, y)$). The consequence (really unnecessary) was the link between intensionality (as the rejection of the reduction to functions of two variables) and externality, although such a link was only a provisional accident from which Russell finally released himself (Grattan-Guinness *1977a*, 95-6, points out the link, but he does not explain it). In a later letter to Jourdain, Russell wrote that in September 1900 he had read Schröder on relations and considered his methods as 'hopeless', whereas Peano provided him with what he was searching for. Russell adds: 'Oddly enough, I was largely guided by the belief that relations must be taken in *intension*, which I have since abandoned, though I have not abandoned the notations which it led me to adopt'.[1] The fact that Russell managed to regard the notation and the conceptual viewpoint as separated things (like in PM) is another sign of his philosophical ballast at this stage.

The application of the mentioned methods gave rise to the first article on this subject: *1901a*. His main achievements are the following: (i) the setting up of a general theory of relations leading to the principle of abstraction; (ii) the first (failed) attempt of a nominal construction of cardinals according to the ideas of Cantor and Peano (see 4.3.1 below); (iii) a theory of series and progressions that would be the general framework of the theory of ordinals (continued in *1902c*) that appeared in POM and PM, which served to insert Cantor into logicism; (iv) a definitive classification of the possibilities to define finite and infinite classes. I shall consider here only the first point, leaving the rest for other sections.

Already in the introduction Russell points out the great simplification of the logic of relations due to Peano's methods, in spite of the lack of making them explicit. That supposed (as we saw above) to give pre-eminence to relations over functions. In §1/*1 Russell starts from one *Pi*, that of *relation* (*rel*), and then he states the convenient notation (*xRy*) and defines the basic concepts. Let us see the definitions (together with the corresponding ones in PM) of the domain (*1.21) and the *relata* of a given term (*1.31) (in PM, respectively, **33.11, 32.131):

$$\rho = x \ni \{\exists y \ni (x\,R\,y)\}\ \ Df; \qquad x \in \rho \,.\, \supset \,.\, \overset{\cup}{\rho}x = y \ni (x\,R\,y)\ \ Df;$$

$$D\,'R = \hat{x}\,\{(\exists y)\,.\, x\,R\,y\}; \qquad \overset{\leftarrow}{R\mathrm{`}x} = \hat{y}\,(x\,R\,y).$$

---

[1] The letter is the one mentioned above from 1910 (in Grattan-Guinness *1977a*, 134). Russell seems to be wrong in the letter when he locates the reading of Schröder in September, since he quoted him in his *1900b* (the paper submitted to the Congress), unless he included some (improbable) revision in proofs.

In the rest of the section Russell states, as *Pp*, that every relation has its converse, that between any pair of individuals there is a relation which does not hold for any other pair, and that the logical sum and product of relations (which are defined) are also relations.

In §1/*2 the relative product (which is a relation as well) is defined. This recourse was destined to play a crucial role in the development of the principle of abstraction and the whole POM and it is an illustration of the importance of the sense of relations:

$$x\,R_1 R_2\, z\ .=.\ \exists y\,`\,(x\,R_1\,y\,.\,y\,R_2\,z)\quad Df.$$

After that, Russell defines the square of a relation ($R^2 = RR$) and, through it, the transitive property ($R^2 \supset R$).

§1/*3 is devoted to membership and §1/*4 to identity and diversity, by means of Schröder's notation (respectively 1' and 0'). The introduction of identity as a *Pi* was important to overcome the former doubts concerning its true status, which did not seem to be that of a relation (especially in *m1900a*) for, if one admitted it as such a relation, it presented *two* terms, which seemed to contradict the idea of *identity*. However, in POM identity does not already appear as *Pi*, although it is maintained as a relation (§95), whereas in PM it is reduced to a mere notation ('I') to abbreviate '$\hat{x}\hat{y}\ (x=y)$' (*50.01).

§1/*5 introduces a new classification of relations which would be crucial for PM. The three resulting types are: many-one, one-many and one-one. The first one is defined as follows:

$$Nc \longrightarrow 1 = Rel \cap R \ni \{x\,R\,y\,.\,x\,R\,z\,.\,\supset_x\,.\,y\,1'\,z\}\quad Df.$$

Finally, §1/*6 states the principle of abstraction, which 'analyzes' equivalence relations as the relative product of a many-one relation by its converse (in 5.2 we shall see it in detail).

The logic of relations appearing in POM consists only of a verbal (non-symbolical) exposition (§§27-30), although a study of the philosophical consequences and applications is added (chapters IX and XXVI). As I shall fully deal in ch. 5 with the ontology of relations, I shall consider here only the two above mentioned problems: the intensional vision and Bradley's argument against relations.

We have already seen that the intensionality of relations seemed to Russell the more suitable alternative with regard to externality. The best proof of this consists in his attribution of the opposite 'philosophical' conviction to those who defended the technical mistake of regarding them as classes of pairs (POM, §27). He explained this mistake through the habit of seeing relational propositions as somewhat more imperfect and unfinished than classical subject-predicate propositions, so that it led to regard relations as classes. Starting from the opposite conviction (which is attributed to Moore) Russell arrives at a more operative treatment, although still with some doubts about its philosophical correctness.

Another reason to consider the intensional viewpoint as the most convenient one was its coincidence with the general pattern of propositional functions, according to which the yielding of classes always depended upon a predicate ($\hat{x}\,(\varphi x)$), and that (as we saw in 4.2.1) led him to see classes rather as class-concepts. Thus, lacking Frege's distinction between meaning and denotation, it was very difficult to overcome the sound argument that two classes can coincide without implying the identification between their class-concepts (e.g. 'man' and 'featherless biped'). Likewise, two relations can have the same extension without being identical (POM, §28). This general parallelism with classes allowed the rejection of the viewpoint according to which relations are to be regarded as classes of pairs. It is true that if we admit this viewpoint it is unnecessary to introduce the idea of relation as a *Pi*, but it is also true that if we want to preserve the 'sense' of relations, the idea itself of ordered pair must be introduced as well as a *Pi*, like in Peano (§98). Consequently, the reasons of conceptual economy cannot be regarded as over-riding. In any case, relations led to the true philosophical axis of POM: the principle of abstraction (see 5.2 below).

The symmetry between relations and classes was consolidated through the distinction between terms and concepts, which was foreign to Moore (POM, §48). By means of it, classes and relations preserve the essence of the concept and they are referred to their respective extensions, the terms. The danger of such a vision is the return it supposes to the subject-predicate pattern, since it is difficult to avoid the impression that the terms are the (logical) subjects and the concepts are the predicates. We could even think that Russell saw the relation between notation and reality as parallel to the relation between predicate (concept) and subject, for although he admits that in the calculus of relations 'it is classes of couples that are relevant', he adds that 'the symbolism deals with them by means of relations' (§98), with which the doubt remains about its character of mere technical recourse. Another sign of the pre-eminence of the subject-predicate pattern consists in the need for emphasizing the corresponding predicates also to form the classes derived from a relation: 'to have a given relation to a given term is a predicate, so that all terms having this relation to this term form a class' (§96). It is true that the notion of the sense of a relation seems to be hardly compatible with its reduction to terms and predicates, but if we remember the attempt to divide any proposition into subject and assertion (see 4.2.1 above), the difficulty decreases.

We come, thus, to the second problem: Bradley's objection against relations. The division into subject and assertion has as a consequence that the latter embraces also the verb, and as verbs are relations, the assertion will embrace both the relation and the adjective (predicate). Relational propositions are not well adapted to this scheme excepting the sense according to which as they are stated between *terms* (subjects) they allow different divisions of a proposition into subject and assertion (e.g. 'A is greater than B' into 'A is greater than' or 'is greater than B'), whereas classical subject-predicate propositions allow only one analysis of this kind. Russell needs to emphasize this difference to defend the irreducible character of relations and, at the same time, he needs to emphasize the parallelism to defend the intensionality. That is why he points out the essential unity of propositions: 'A proposition, in fact, is essentially a unity, and when analysis has destroyed the unity, no enumeration of constituents will restore the proposition' (POM, §54). However, the problem of where to locate this unity immediately appears. If we locate it in the verb (the relation), the intension prevails and the propositions as mere complex concepts (like in Moore) vanish; and if, on the contrary, we consider the verb as one more term, then the proposition as a unity vanishes, unless we introduce other 'relations' between the remaining terms, with which we would arrive, again, at Bradley's objection. (It is the problem of the relations in themselves and the 'relating' relations, which led Russell to the difficult view that relations have no particular instances; see 5.4 below.)

In *m1899a* this problem was already considered as the greatest difficulty of the theory of relations. If a relation relates two terms, it seems that the relation itself must, again, be related with them. As relational propositions are ultimate, the endless regress seems unavoidable. Russell's defence consists in the distinction between two kinds of endless regresses (POM, §§55, 99): according to one of them, which can be admitted, the infinite implications of a proposition are not a part of it, so that they do not threaten its meaning; the other type, on the contrary, would transform relational propositions into nonsense if we admit that the relation between the relation and the related terms is a part of its meaning. Russell's position consists in denying this by arguing that 'a relating relation is distinguished from a relation in itself by the indefinable element of assertion which distinguished a proposition from a concept' (§99). It was an efficient position to classify this endless regress in the class of the harmless ones, in resting on the unity of the proposition, which does not admit strange additions (as would be the added relations). However it gives rise to the difficulty of being an attack against Moore's theory of judgment, which must have avoided the distinction between propositions and concepts with the argument (later regarded as proceeding from logical atomism) that there is nothing in the proposition not being in its constituent elements. Of course this depends, again, upon the belief (later discarded) that any word has some meaning.

Summing up: the defence of the ultimate character of relations leads Russell even to break with Moore's philosophical framework, which was precisely the cause of this same ultimate character. The compromise between intensionality and externality was, however, abandoned starting from the contact with Frege (already in one of the appendices to POM, §485). The extensional position of PM, where only the notation was preserved, avoided such philosophical problems by reducing relations to propositional functions of two variables.

4.3. Arithmetic

4.3.1. The definition of cardinal number

It is usually accepted that, a little after attending the Paris Congress of 1900, Russell obtained (independently from Frege) his famous definition of cardinal number as a class of classes for the first time. In the same way, it is believed, this definition appeared in the famous essay we considered in the former section (*1901a*). Using this definition and similar others, Russell conceived his logicist construction of mathematics in contrast to Peano's construction (and Dedekind's), based on axioms (*Pp*) and indefinable ideas (*Pi*). The received view adds that the role played by Peano was limited to providing notations and ideas to facilitate Russell's achievement. Russell himself has contributed to this view about his role in the history of foundations. Thus, we read in Marsh's introduction to *1901a*: 'Asked what idea in the paper he now feels to be the most important, Russell replied "my definition of cardinal number" —which here appeared in print for the first time' (*1901a*, 1). Likewise, we read in IMP on Frege *1884a*: 'the definition of number which it contains remained practically unknown until it was rediscovered by the present author in 1901' (p. 11; we find the same in MPD, 54).

However, as we have seen it was the contact with Peano and his school that gave Russell *all the ingredients* (including reasons to accept Cantor's transfinites) needed to succeed in logically defining cardinal number and to obtain a general logicist construction of mathematics. Firstly, Cantor provided the means to define number before the distinction between finite and infinite, as well as the technical recourses to present arithmetical operations through logical relations. Moreover, he gave a logical definition of infinity (closely related to those by Bolzano and Dedekind) by means of the correspondence between classes (see 1.2 above). Secondly, despite the fact that Peano was not a logicist (see Grattan-Guinness *1980a*, 87), he clearly saw the possibility of constructing a logicist arithmetic through Cantor's concept of correspondence. Besides, he gave logicist definitions of particular cardinal numbers as well as made use (trying to overcome it) of Pasch's construction of real numbers as segments of rationals (just the one chosen by Russell; see 3.2.3 and 3.2.4 above, and 4.3.6 below).

Thirdly, Pieri invented a method to transform definitions by abstraction into nominal definitions and constructed a geometry very close to logicism including *order* as a main notion, and also a logical definition of the straight line (both recourses were fundamental for Russell in POM). Finally, Burali-Forti applied the idea from Pieri to obtain nominal definitions of the three arithmetical primitive ('indefinable') notions of Peano's *Formulaire* in logical terms of *class* and *relation* (which were used by Peano in proposing his clear logicist definition of cardinal number in 1901). He also introduced a philosophical justification of nominal definitions and invented a method to transform definitions 'by postulates' into nominal definitions: to present the postulates (axioms) as *properties* of a general concept (which was widely applied by Russell in his writings) (see 3.5 above and 4.4 below). However, Russell's principle of abstraction was never explicitly used by Peano or his followers, either to justify definitions by abstraction or to justify nominal definitions.

The ultimate priority for a nominal logicist definition of cardinal number is due, of course, to Frege *1884a*; but as we saw in 3.5 not even the Italian authors became completely aware of his definition despite their being familiar with some of his writings (especially Peano). The truth is that Russell defined for the first time cardinal number as a class of classes in *1901a* (his first publication in Peano's review) without mentioning previous writings by Pieri and Burali-Forti. Moreover, Peano's *Formulaire* of 1901 (whose preface was dated as of January) preceded the publication of Russell's paper (in the middle of 1901), and we saw how that edition of Peano's celebrated work contains the famous definition (see 3.2.4 above).

It is possible to argue that Peano's definition was implicitly referring to Russell's paper, which was waiting for publication in his hands. The grounds for this supposition are in Russell's allusions (AB1, 145; MPD, 55) to September 1900. Thus Kennedy (*1973b*, 369) writes: 'Peano's remark was almost certainly prompted by Russell's definition of cardinal number as a class of classes', and he adds that although Peano *1901a* was published before Russell's paper, this one was known to Peano since October. We shall see, however, that the supposed definition does not appear symbolically in Russell *1901a*. In fact, we only find a *verbal* version of it at the end of §1 and an obscure later reference (also verbal). We must add that *1901a* includes quotations from Peano *1901a* and that the letter from Peano accepting the paper for publication was dated March 1901 (Kennedy *1975a*, 206). These facts suggest the idea that *perhaps* Russell modified his first version when he knew Peano *1901a* at the time of composing the final version (including the correction of proofs), and that perhaps one of these changes was the addition of the verbal definition mentioned above. Finally: only in Russell's publications from 1902 did the symbolical version appear. We shall now see that a detailed examination of Russell's paper and a comparison to the original manuscripts confirm my interpretation.

Russell *1901a*, §2 is devoted to cardinal numbers. However, at the end of §1 he pointed out the problem in writing that definitions by abstraction define a class and not an individual. The solution is given by the principle of abstraction analyzing a transitive and symmetrical relation (R) as the product of a many-one relation (S) and its converse. Then we can take the class of *relata* (of R) for a given term $(\breve{\rho}\, x)$ 'as the individual indicated by the definition by abstraction; thus for example the cardinal number of a class u will be the class of classes similar to u' (*1901a*, 10). Curiously, *this is the only place where this definition appears so clearly* (neither verbally nor symbolically), including of course the entire section §2 (entitled 'Cardinal numbers'). In fact the general idea recalls the way followed by Burali-Forti (see 4.5.2 above).

§2 begins by defining similarity between classes as a one-one relation such that one of the classes is included in the domain and the other in the converse domain (*1.1). (The same definition was maintained until PM with only notational improvements.) After that, the similarity is stated as an equivalence relation, and then the principle of abstraction is applied to demonstrate the existence of the relation S and the corresponding relative product (*1.3):

$$\exists Nc \longrightarrow 1 \cap S \ni (sim = S\breve{S}\,),$$

and to define the class S of these relations (*1.4):

$$S \;=\; Nc \longrightarrow 1 \cap S \ni (sim = S\breve{S}\,) \;\; Df.$$

At this point the obscure additional reference mentioned above appears. Thus, the definition of cardinal number 'as a class of classes, of which, each has a one-one correspondence with the class "cardinal number"' is regarded as a definition 'by abstraction' (*1901a*, 11). We can easily confirm the unclearness to distinguish between: (i) the two kinds of definition (by abstraction and nominal), and (ii) the number of a class, the general concept of number and the class of numbers (in PM clearly distinguished as $Nc'\alpha$, Nc and NC; see below).

It might seem that the symbolical definition quoted is referring to the class of cardinal numbers, since there is no other definition playing this role. However, the entity actually defined is rather a class of relations; in particular the class of the many-one relations S such that the relative product $S\breve{S}$ is equal to the similarity relation. (We shall see below —in 5.2— that any product of a relation and its converse is generally an equivalence relation, so that the mentioned definition might perhaps be regarded as pointing out simply the class of equivalence relations.) In any case Russell continues with the same class in the next step, although now in terms of the domains (i.e. classes and not relations). In fact Russell chooses σ (the domain of one of the relations S of the class of relations S) as (it seems) Nc and then he demonstrated that (i) $\sigma = Cls$ (*1,5), (ii) all classes constituting such domains are similar to each other (*1.52), (iii) all classes similar to one of them (to one of these domains) belong to this class of classes (*1.54).

A sign that such domains would be regarded as cardinal numbers could be the following logicist assertion (*1901a*, 11):

> all classes which form the domains of different relations of the class S are similar (sim), and... all classes similar to one of them belong to this class of classes. The arithmetic of cardinal numbers applies in its entirety to each of these classes.

For the same reason, the next Peanian definitions of 0 and 1 (*2.4.7) show his respective *definienda* written as 0_σ and 1_σ. At last, Russell chose the Cantorian (and Italian) way to define inequality (without mentioning Cantor) in terms of (i) non-similarity between classes, and (ii) similarity between the lesser class and a subclass of the greater one (*2.1). In this same line the addition is defined in terms of class union (*4.1).

At this point the essential elements of the cardinal number construction are completed. We have seen how the use of the principle of abstraction is unclear from the philosophical point of view. Besides, the non-use of Peano's handy '*Num a*' notation produces a strange impression. In any case, the definition of cardinal number is formulated only in an obscure way and, of course, the celebrated *symbolical* definition in terms of a class of classes is missing. It even seems that Russell never thought of including a nominal explicit definition of the number of a class in logicist terms at the moment of composing the article.

Only the recourse to the unpublished manuscripts throws some light on these problems. There are two relevant manuscripts. The second one (*m1901?*; a French version corresponding to the final actually published paper) remains incomplete, but it contains the sections we need. If we compare it to *1901a*, two notable facts arise. First, the quotations from Peano's *Formulaire* of

1901 *are already present*, therefore when Russell wrote this version he had seen this work (which, as we saw, contains the symbolical logicist definition of *'Num a'*). This is important because it would be possible to think that these quotations were added in proofs. Second, the only clear (although verbal) definition of the number of a class as a class of classes (mentioned above as appearing in the end of §1 in *1901a*) is *completely missing*, therefore it was *added*, very probably in proofs. In any case Peano's definition of *1901a* was available to him. The rest of the manuscript presents no more differences.

Consequently, we can confirm our hypothesis according to which the section on cardinals (§2) in *1901a* was not conceived to give a nominal definition of the cardinal number in terms of a class of classes. Moreover, definitions by abstraction are not clearly distinguished from nominal definitions in practice. It is true that the definition of S is nominal, but of course it is not possible to identify this entity with the one referred to by Russell in the *missing* addition (the number of a class).

A study of the first manuscript (*m1900c*; the English original version) produces even more interesting results. It includes some material on geometry to be discarded in the final French version as well as definitions and demonstrations modified afterwards. On the title page we read 'Oct. 1900'; therefore this was the version Russell had in mind in his autobiographical writings (see above) when he wrote 'September', and *not the version actually published*, whose date is not known but (as we have seen) must be later than that of Peano *1901a*. In this version we find the same fact: the supposed attempt to define cardinal number as a class of classes is lacking.

I shall consider only two points from its content. First, there is the preface: after stating the need for making relations explicit (in the line of Moore and the manuscripts from 1898-1900), Russell adds that only this way allows 'a simplification and a generalization of many mathematical theories', although he has not yet said anything about the link between the possibilities offered by relations and nominal definitions. However, immediately after that we find two lines *clearly added later* mentioning this connection: 'and it enables us to render all definitions *nominal*', and a reference to Burali-Forti *1900a*. By contrast, the published version preserves the stated link but *omits* the reference to Burali-Forti (Russell *1901a*, 4). Hence we cannot avoid concluding, at least, that (i) Russell did not try to find nominal definitions *when he composed the first version*; (ii) the reading of Burali-Forti (especially his logicist emphasis to *reduce* arithmetical notions to logical terms) was almost certainly the reason to state the mentioned connection. These conclusions have probably some relation to the *addition* I mentioned above concerning the verbal definition of the number of a class in *1901a*.

Second, we also have the content of §3 on cardinal numbers. If we compare it to the published version, our general interpretation seems to be also confirmed. The beginning is the same: the similarity between classes is defined, this similarity is stated as an equivalence relation, and the principle of abstraction is then applied (although here without demonstration; see 5.2 below). However the principle is described as *presupposed* in definitions by abstraction in a way that 'states that, in all cases where definition by abstraction is formally allowable, there exists the entity which is defined' (*m1900c*, 7). Therefore the principle is not presented as a recourse making definitions by abstraction unnecessary, as in the customary way of later writings. It seems, rather, that the principle was thought *to strengthen the method of abstraction and not to dispense with it*. This impression is confirmed when we continue the comparison, for after (*1.3):

$$\exists Nc \longleftrightarrow 1 \cap S \ni (\text{sim} = S\breve{S}),$$

the class of relations S is not introduced, as in the published version, but instead the following definition is given ((*1.4):

$$\text{sim} = S\breve{S} \,.\, \supset \,.\, Nc = \breve{\sigma} \quad Df.,$$

which is lacking in *1901a*. Now the cardinal number (Nc) is identified directly with the converse domain of one of the relations S ($\breve{\sigma}$), and then we find the cardinal number of a class defined (it seems) as the class corresponding to the relation S (*1.5):

$$\text{sim} = S\breve{S} \,.\, u \in Cls \,.\, \supset \,.\, Nc\text{'}u = \imath u_s \quad Df.$$

The final part is very like the published version. We find $\sigma = Cls$; the definition of inequality in Cantor's terms; and the customary definitions of 0 and 1, although now replacing σ for Nc (on the apparent role of σ in *1901a* see above).

Thus, the introduction of the cardinal number by means of the domains (the domain and the converse domain) of a relation S is clear. However, the final way (in the published version) was to eliminate this definition and to preserve only the class of relations itself, avoiding to identify Nc with σ but (it seems) making σ play the role of cardinal number (as we saw above). I am under the impression that Russell was avoiding offering a *definition* of number, which can be interpreted, together with the former presentation of the principle of abstraction as a *justification* of Peano's methods based on abstraction, as an attempt to preserve these methods through a definition by abstraction of cardinal number. In this way the principle of abstraction would permit one all the necessary elements to secure the existence of the entities defined by abstraction (as Russell seems to say in *1901a*; see above) and to avoid any attempt to nominally define Nc.

A note clearly added later in the same manuscript (next to definition *1.4 quoted above) says: 'This Df wont do: there may be many such relations as S. Nc must be indefinable' (*m1900c*, 11). With the added note Russell seems to become aware that in setting $Nc = \breve{\sigma}$ he was pointing out a relation S that really is distinguished only by its belonging to a class that in the published version appears as S. Besides, he seems also to become aware that by doing so he was affecting the supposed undefinability (in logical terms) of Nc, one of the essential doctrines inherited from Peano, which coincided with Russell's view as expressed in the manuscripts of 1898-1900 (see 2.2 and 2.3 above).

4. The principles of mathematics

The reading of Burali-Forti (who had already constructed nominal definitions of Peano's arithmetical 'indefinables' in terms of class and similarity) seems to have been one reason that Russell became aware (better than Burali-Forti himself) that some logic of relations could reach, not only the logical definition of *Nc*, but a whole programme of transforming definitions by abstraction into nominal definitions. This method concided at least with another of his fundamental goals: to lead the *analysis* (in Moore's sense) to its ultimate consequences; and this was what he tried to do in the published version, which, although does not yet contain the standard definition of the cardinal number of a class, it includes some criticisms of the ambiguity of definitions by abstraction (in the same line of Burali-Forti) and, for the same reason, a new interpretation of the principle of abstraction (although still not in the standard terms according to which this principle must dispense with these definitions).

The later publications develop this line towards a greater clarity. The standard symbolical definition, identical to that of Peano *1901a*, appears only in the articles from 1902. Both *1902b* and *1902c* already offer clear definitions of the number of a class and of *Nc*. In *1902b* we find (*u* and *v* being classes, and 'μ' a notation for 'number of') (*1.2 and *1.3):

$$\mu u = Cls \cap v \ni (u \, sim \, v) \quad Df.;$$
$$Nc = Cls^2 \cap z \ni \{\exists cls \cap u \ni (z = \mu u)\} \quad Df.$$

The second of these definitions improves upon the corresponding one appearing in *1902c* (*7.11) and is very similar to that by Frege in presenting the cardinal number as being the number of some class (the same idea of Burali-Forti *1900a*, 303). However, these definitions by no means use the principle of abstraction; in fact it is only mentioned (see 5.2 below). On the contrary, the principle appears again in POM, where it plays the role of a true foundation and appears in almost all the constructions. In particular, in the case of the cardinal numbers, the construction is presented (now in an explicit way) as the connection between logic and arithmetic (although the definitions are offered only in verbal form).

Moreover, Russell's presentation of the principle in POM proposed it as a recourse to overcome the supposed ambiguity of the corresponding definition by abstraction (POM, §111), identifying the number of a class with the class of all classes similar to the given class. Thus the membership to this class of classes is equivalent to the property common to all similar classes. The supposed ambiguity between the class and the predicate (or class-concept) is solved by Russell pointing out the symmetry between the predicates 'similar to *u*' and 'similar to *v*'; for both lead us to the same definition, which defines the class itself and not the class-concept. At last Russell states: 'this definition allows the deduction of all the usual properties of numbers, whether finite and infinite' (§111). The logical features of this definition allow as well the rejection of any possible predicate common to all such similar classes (from the 'mathematical' viewpoint). Summing up: the function of the principle consists in replacing the supposed common property inferred from the equivalence relation for the class of terms having the given relation to a given term. (In 5.2 below we shall see that there are really *two* very different relations; when they are distinguished and analyzed the principle seems to lose its power as a foundation.) Thus, the role of the principle is now very different to that which it played in the first writings inspired by Peano, where it seems to be destined to *justify* definitions by abstraction.

In PM (and earlier in *1908a*) the construction of cardinal numbers ignores again the principle of abstraction (it is only mentioned; see 5.2. below). The offered definitions (already definitive) contain improvements making the concept of number to resemble Frege's ideas (especially when Nc is presented as a relation). The definition of the number of a class presents, however, no change (*100.1):

$$Nc'\alpha = \hat{\beta} \ (\beta \ sm \ \alpha),$$

but, regarding Nc as a relation, it is equal to $\overrightarrow{sm}$ (classes similar to a given class), therefore $Nc = \overrightarrow{sm}$ (*100.01). The class of cardinal numbers will be (PM, II, p. 5):

$$NC = \hat{\mu} \{(\exists \alpha) . \mu = Nc'\alpha\},$$

namely, the class of numbers being the number of some class. However, by regarding the last *definiens* as the domain of the relation Nc, we have (*100.02):

$$NC = D'Nc \ Df.$$

The actual non-use of the principle of abstraction (as a ground) in the works mentioned (excepting in POM where it played an important but 'philosophical' role) was never justified for by Russell. As we shall see in 5.2 the exposition of the principle was modified in later writings, and the alterations, although seemingly unimportant, seem to hide the discovery of its uselessness.

4.3.2. Finite and infinite

Starting from the Congress of 1900 Russell had already abandoned all of his doubts about the infinite, taking advantage of the possibilities of Peano's notation to offer his first symbolical definitions of such a concept. In *1901a* the celebrated Cantorian definition already appears in terms of similarity of a class to one of its proper parts (§4, *3.51), which had been given by Peano in the *Formulaire*. However, Russell proposes and makes use of a different definition which he regarded as more fruitful from the deductive viewpoint: a class is infinite if it remains similar to itself when one subtracts a term from it (*1901a*, §4, *1.1). Then the finite class is defined as the non-infinite one and the unit class ('*Elm*' following Padoa) as a finite class (for it does not remain similar to itself when one subtracts a term from it), in the same way that the null class (where the substraction of a term is impossible). The possibility of defining finite numbers by mathematical induction (and therefore also the infinite ones) is also discarded by deductive reasons.[1] Thus the independent status of the two discarded definitions, the Cantorian one and that based on mathematical induction, remains requiring clarification.

In *1902b* Russell solves the problem by adopting the discarded possibility (see Couturat *1904a*, II, 227). Thus, in deductively arriving at the Cantorian definitions, he demonstrates its equivalence with the 'ordinal' definition. To do that, he firstly defines the finite cardinal numbers as those that can be obtained through mathematical induction (following Burali-Forti), and the infinite ones as the non-finite (*1.8). Then he defines Cantor's *aleph* ($\aleph$) as the cardinal belonging to the (infinite) class of the finite cardinals, and he demonstrates that it is included in every infinite class. Finally he demonstrates the usual Cantorian properties of the finite and infinite classes in terms of similarity (*2.7.71):

$$u \in cls\ infin\ .\ v \in cls\ fin\ .\ v \supset u\ .\ \supset\ .\ u\ sim\ u - v;$$
$$\alpha \in Nc\ infin\ .\ n \in Nc\ fin\ .\ \supset\ .\ \alpha - n = \alpha = \alpha + n;$$

which directly lead him to the definitions of the finite and infinite cardinals starting from the similarity (or not) with a proper part (**2.75.76):

$$Nc\ fin = Nc \cap n \ni (n - = n + 1);\qquad Nc\ infin = Nc \cap \alpha \ni (\alpha = \alpha - 1),$$

with which the two former definitions cease to be independent.

In POM (§§117-9) the presentation from *1901a* is reproduced since the definition based on induction appears only as a possibility,[2] although Russell concludes that it is equivalent to Cantor's definition in the sense that the former can be regarded as the consequence of the latter. Russell claimed later on that such an equivalence presupposed the multiplicative axiom, but the important thing here is that this equivalence provided the construction of the infinite with a great generality: it allowed the unity of infinite ordinals and cardinals (for they do not fulfil mathematical induction) as well as the distinction between them (for only the latter fulfil the standard definition of whole and part) (POM, §250). In any case the acceptance itself of the definitions shows the pre-eminence of the constructive viewpoint over the philosophical one followed in the manuscripts from 1898-1900.

However, in PM the two definitions are quite separated, since the first method gives rise to inductive and non-inductive classes and numbers, whereas the second yields non-reflexive and

---

[1] In his *m1900c* (the ground for this work I studied above) he offered only a definition of finite numbers as those cardinals changing when we add 1 to them (p. 14). On the other hand, infinite numbers are not defined (perhaps Russell still had some doubts concerning them).

[2] It is possible that the composition of *1902b* was later to POM, for it appeared in 1902, whereas the manuscript of POM was finished in May 1902. Kenneth Blackwell thinks (personal communication) that this article was written in June, although he has not yet found a definite evidence.

reflexive classes and numbers (PM, II, 181 ff). The *inductive* cardinals are those that can be reached, starting from 0, by successive additions of 1, i.e. those constituting the *posterity* of 0 with regard to the *hereditary* relation $(+_c 1)*$. In symbols (*120.01):

$$NC\ induct\ =\ \hat{\alpha}\ [\alpha\ (+_c 1)* 0]\quad Df.$$

This definition has the consequence that mathematical induction behaves as a defining feature rather than as a principle (although the *existence* of cardinals which do not fulfil it requires the infinity axiom). The *reflexive* cardinals are introduced starting from the corresponding class, which is defined as that maintaining a one-one relation to a proper part, i.e. a class equal to the domain of this relation and containing the converse domain as a proper part. A reflexive cardinal will be precisely the cardinal number of such reflexive classes (*124.02):

$$NC\ refl\ =\ N_0 c\ ``\ Cls\ refl.$$

As for the equivalence, since any inductive class is greater than a proper part of itself, any inductive cardinal will be non-reflexive as well (although to prove that any non-inductive number is reflexive requires, according to Russell, the multiplicative axiom) (PM, II, 184).

4.3.3. Quantity

As I showed in 2.3 Russell reduced, in the manuscripts, magnitude into two indefinables (the 'greater' relation and every particular magnitude). Thus a *quantity* was any term with magnitude, so that symmetrical relations were eliminated in eluding the supposed 'common properties' of equal quantities. This supposed an implicit use of the principle of abstraction, for equality is defined as 'sameness of magnitude' instead of considering it as indefinable and using it to define magnitude as the common property of equal quantities. The 'logic' of the external relations reduced those common properties to independent terms and to the corresponding asymmetrical relations (the only 'genuine' ones). This procedure fulfilled the requirements of a pluralist philosophy in spite of increasing the number of indefinables.

The contact with Peano supposed the discovery that the logical recourses destined to give an account of real numbers could be used to explain the concept of quantity. In fact this subject appeared only in the first edition of Peano's *Formulaire* (see Couturat *1904a*, IV-V, 675 ff). In practice, as Russell already had an almost finished version of POM1 available, this viewpoint had supposed the need to discard an entire part (that devoted to quantity), which constituted a great opportunity to show the usefulness of Moore's philosophy, as well as an authentic exhibition of the possibilities of relations to solve the traditional problem of the connection between quantity and number (which was present in Russell from FG). The solution consisted in preserving all the written material (the published part III of POM) adding some little details.

In a new introduction (POM, §§149-50) Russell explains how the new arithmetic (Weierstrass and Cantor) had reached the independence from quantity by reducing it to order, continuity and real numbers (in fact Couturat had already said that in his *1898b*). In the last analysis, however, Russell denies that quantity is definable in terms of logical constants, so that it would not belong to pure mathematics (nevertheless Burali-Forti had already reduced quantity into logical terms in his *1899a*; see 3.5 above). Thus the only possible way to justify the maintaining of the four following dense chapters was an effort to destroy the traditional view which identified arithmetic and quantity (which at that moment was already completely discredited). Even the view according to which the problem could be reduced to two basic theories (the relative and the absolute ones) was nothing but an attempt to take advantage of the material included in his *1900b* and the improved version *1901c* (although there he used to speak rather of 'position'; see 4.1 above and my *1987a*, 361 ff). The novelty now is the explicit use of the principle of abstraction.

The essentials of the relative theory (POM, §154) replace magnitude for equality and applies inequality to quantities, with which equal quantities have nothing in common that unequal quantities do not have. Thus magnitude is dissolved into mere relations. On the other hand, the absolute theory (§155) attributes equality to quantities and inequality to magnitudes, so that the common property of equal quantities is precisely the magnitude itself. The advantage of the latter theory is that the number of axioms ruling it is lesser, since inequality disappears. Russell applies the new recourses by defining magnitude as the class forming the field (the domain plus the converse domain) of the inequality relation, but this adds nothing to the standard former position.

The choice between the two theories is automatic through the principle of abstraction (POM, §157), which solves the equality between two terms into a 'special' common relation with a third term (and not into a common property). The problem with quantities is more complicated than with positions because not only quantities point out a particular magnitude, but magnitudes themselves have to come together into 'types' (only then it is possible to compare two of them), with which a new relation (as well 'peculiar') appears between every magnitude and the corresponding quality. As we shall see in 5.2 the problem can be simplified by distinguishing between the two involved relations.

In PM quantity in the philosophical sense does not appear. However, the proof that Peano's view (mentioned above) was the correct one lies in the fact that the title *Quantity* (of the last part of the work) corresponds to a part where the generalization of number until reaching the reals is stated and then they are applied to measurement. Magnitude and quantity are used as synonymous and they are defined as operations or relations (in particular as families of vectors) each one of which is a kind of quantity with special properties (PM, III, 233, 339 ff). In any case, the principle of abstraction does not appear.

4.3.4. Order

Concerning this subject we can also notice a great continuity with regard to the stage previous to the Congress of 1900. The central ideas of the manuscripts (see 2.3 above) tend to be preserved under the philosophical essential viewpoint, although they are technically improved. In 1901 Russell published two articles related to order. One of them (*1901c*) has already been considered in 4.1.2 as a sign of the changes after the Congress. The other (*1901b*) is especially interesting because, although a first version was already read in May 1900[1] (and thus it supposed as well a consequence of the manuscripts position), it incorporated some novelties when published. It reproduces the traditional ideas, but includes a set of different ways to yield order (taken from Vivanti and Vailati, who are mentioned), which anticipate the POM treatment, and a rejection of *logical order* (which was so important in the former period) based on a certain progress concerning the analysis of implication. According to this analysis, implication cannot be clearly distinguished from equivalence: 'a relation which is sometimes symmetrical and sometimes asymmetrical seems intolerable, and should be avoided if possible' (*1901b*, 36-7). Here Peano's logic definitively clarified the problem by restricting implication only to propositions.

Finally, a short discussion of several applications of the stated thesis (the reduction to an asymmetrical relation) is added, although with no use of the principle of abstraction (perhaps because of a limitation to the establishment itself of order rather than an identification of some 'absolute' feature of its essence, which would consist in the 'externality' of the terms involved). In any case the basic idea in POM (and completely new in relation to the former manuscripts) remains clarified: the mechanism of the correlation of series always gives rise to absolute (independent) series, which can be considered as a starting-point to state relative (by correlation) series (*1901b*, 31). This leads Russell to set an ontological link between order and logic: 'to be an independent series is to have a distinguished place among entities. This fact explains why such series have great philosophical importance, and why the theory of order is one of the most essential parts of logic' (*ibidem*).

The articles published in Peano's review (*1901a* and *1902c*) are also important for the theory of order. The first one because it reaches series and progressions *starting from* relations (already from a symbolical viewpoint). The second because it completes the exposition of the subject with the symbolical transcription of the second part of Cantor *1895a* (including some corrections of the first version). As all this has a clear relation to ordinals, I shall return to them in 4.3.5.

I come now to POM, where the most important thing will be to understand the way in which the technical theory of relations is applied until pointing out the principle of abstraction as a philosophical foundation. The undisputed starting-point is that 'order' is equivalent to 'series', so

---

[1] The information concerning the previous reading proceeds from a personal communication of Kenneth Blackwell. This work is really an odd hybrid still containing some pre-logicist ballast. It seems to be, then, another instance of taking advantage of the material previously written. Curiously it was published in the issue of January in *Mind*, precisely when Russell was composing his *1901d*, a true logicist manifesto (see 4.1.2 above).

that, before analyzing order itself, it is offered an inventory (similar to that from *1901b*) of the six possible methods of generating series (§§189-94). I only have room for a small reference to them (Couturat *1904a*, II, 230 ff, offers an excellent summary). The first method applies an asymmetrical and intransitive relation to a collection so that the relation and its converse take place only once between every two terms. In the second one the relation (*P*: precedes) is also asymmetrical, but transitive, which allows no consecutive terms (although closed series are impossible); thus any two terms will be xPy or yPx (the same for the converse relation $\breve{P}$) and for any term x the rest of them will be divided into two classes (according to xPy or zPx). If then we take two terms xPy, the rest will belong, either to πx and πy (the referents of x and y), or to $\breve{\pi} y$ and $\breve{\pi} x$, or to $\breve{\pi} x$ but not to πy. Therefore, if we take three terms, one of them will always be *between* the other two, i.e. the series will be single (this would be the method finally chosen). The third consists in ordering the terms by means of the asymmetrical relation of its distance to one given term (which can be reduced to the former one). The fourth method is based upon the triangular relations (like *between*), and the last two try to generate continuous and closed series through the application of an asymmetrical relation to a class of relations (the fifth) and the four-term relation of separation (the sixth).

The next step will be to find out what all these methods have in common and, once this element is discovered, to start the relevant logical analysis (POM, §195). As all of them make use of either the relation *between* (which requires three terms) or the relation of *separation* (which requires four), the analysis has to be applied to every one of them (although the first case will be the most important).

Russell starts from the second method, which is based on an asymmetrical and transitive relation (POM, §§196-202). According to it, '*y* is between *x* and *z*' (in short '*xyz*'; $\bar{R}$ being the negation of *R*) will be defined in the following way (and will give rise to two axioms as consequences) (§197):

$$xyz = xRy \ . \ yRz \ . \ y\bar{R}x \ . \ z\bar{R}y;$$
$$(\alpha) \ xyz \ . \ yzw \ . \supset . \ xyw;$$
$$(\beta) \ xyz \ . \ xwy \ . \supset . \ wyz.$$

The first axiom requires a transitive relation, but if we would have an intransitive one *R* (like in the first method) we could replace it for another *R'* (some positive power of *R*) in a way that, if *xyz* holds for some power of *R* (as being the case with intransitive relations), then it also holds for *R'*, with the only condition that none of such powers are equivalent to $\breve{R}$ (which will transform the series into a closed one, and *between* cannot be applied to these series). As for the second axiom, it is only possible if the relation is not a one-one relation (since *xRy* and *xRw*). Consequently, the two axioms are consequences of the definition.

Before giving a final result Russell considers some philosophical objections that, in general, he solves by pointing out the pre-eminence of terms over relations as the objects of order (POM,

§201), but with that he emphasizes the distinction between terms and concepts, contrary to the spirit of Moore's philosophy. Besides, he considers Bradley's objections against relations and defends the definability of *between* (§202), but I shall examine both arguments in 5.4.

The definitive formulation of order is the following: 'A term y is between two terms y and z with reference to a transitive asymmetrical relation R when xRy and yRz' (POM, §202). This will be the necessary and sufficient condition and, and at the same time, the 'true meaning' of *betweenness*. The conclusion can be objected to with the argument that although the six considered methods form an inductive ground to the later essentialist (Socratic) analysis, however Russell forgot to take into account the non-constructive usages of order, i.e. those appearing in ordinary language (to put a set of things in order, etc.). That is why Russell's essentialism, in prejudging the admissible constructions without demonstrating that they exhaust the possible usages, seems incomplete. This remark is important because it shows the way in which Russell moves away from Moore's analytic practice which, although also rejected the mere lexical analysis, always considered it as a fundamental criterion (see 1.4 above). The criticism is also relevant in so far as Russell began his whole analysis by proclaiming that he was looking for the *logical constituents* of order (§194) and that this analysis will be 'of purely philosophical interest, and might be wholly omitted in a mathematical treatment of the subject' (§195).

Concerning the relation of *separation* among four terms (the second type to analyze), it is also reduced to an asymmetrical and transitive relation (POM, §204), which constitutes its necessary and sufficient condition. Sufficient because when the latter relation takes place among a series of four terms, the former also appears; necessary because (here Russell mentions Vailati) when the relation of separation takes place, it can be reduced (at least formally) to the existence of an asymmetrical and transitive relation between every two of the terms (e.g. the relation P in the second of the methods to yield order). The identification that this supposes between open and closed series (between linear and circular order) is then transformed into something even previous to the reduction itself to asymmetrical relations, for it makes it possible for a closed series to become an open one with only fixing a first term. Besides, as the order has the pre-eminence, a relation between projective geometry and arithmetic is stated where the open series had the pre-eminence (from Cantor). That is why it is difficult to understand Russell's insistence on the assertion that all series are open from the mathematical viewpoint (§224) and that there would subsist a philosophical difference with regard to the generating relation (§228), for if the distinction was technically justified by considering the need to maintain certain terms fixed in the closed series, the important philosophical thing to emphasize would be the similarities.

It is sufficient to point out here the way in which Russell needs to emphasize terms to overcome Bradley's objections and, at the same time, relations to defend pluralism. On the other hand, he also needs to maintain relations to replace properties (which vanish, according to the principle of abstraction, into other terms and a certain relation), although sometimes he also reduces them to concepts (predicates or adjectives) in spite of Moore's prohibition to do so. In arriving at geometry, we shall see how relations return to obtain the pre-eminence and how points (the terms of geometry) remain in such a situation that they are neither defined (for they cannot be constructed; at least at this stage of Russell's evolution) nor admitted as indefinable (for it would be contrary to logicism); and this in spite of the fact that in chapter 51 of POM Russell defended points for philosophical reasons related to the need for avoiding idealism.

PM supposed the definitive version. Here we do not find *order* defined, but the relation itself yielding series, i.e. the *serial relation*, which is contained in diversity (it is more or less equivalent to asymmetry), transitive and connected (properties already given in *1902c*). In general we have (*204.01):

$$Ser = Rl`J \cap trans \cap connex \quad Df.$$

In more detailed terms and *P* being the considered relation (*204.11):

$$P \in Ser \ . \equiv : P \subset J \ . \ P^2 \subset P : x \in C`P \ . \supset_x . \ \overrightarrow{P}`x \cup \iota`x \cup \overleftarrow{P}`x = C`P,$$

which expresses the connected property (that the relation or its converse take place between every two terms of the field) in terms of the logical sum of the referents and the relata of any term and this same term. The problem of the repeated terms is solved like in the old theory of the absolute position (although now without the principle of abstraction): the series to which they belong are series by correlation to genuine series, which fulfil the condition $P \subset J$ (PM, II, 497). Finally, the pre-eminence of relations over terms is stated by identifying the series with the generating relation: 'The generating relation determines the order, and also the class of terms ordered, since this class is the field of the generating relation. Hence the generating relation completely determines the series, and may, for all mathematical purposes, be taken to *be* the series' (PM, II, 498). This is the victory of constructive (mathematical) definitions over philosophical ones (the analysis into indefinable terms), although Russell not always distinguished clearly between the two types (see particularly 5.3.1 and 5.3.2).

4.3.5. Ordinal numbers

Russell's construction of the ordinals after the Paris Congress is also very different to the position of the manuscripts. However, the independence of order with regard to number is preserved (the order was also defined there by means of relations having nothing to do with quantity). The priority over cardinals is also preserved, but as the negation of such a priority is only attributed to the traditional association between ordinals and the process of counting (which is rejected by Russell), he does not delve deep into the reasons related to conceptual economy, which would be decisive in POM (although with some doubts). The added things are, then, the most important ones: the application of the theory of relations; the definition of progressions as special series (starting from Cantor and Peano); the likeness[1] between relations (and the pre-eminence of abstraction); and finally the theory of relation-numbers, constructed through the order types corresponding to well ordered sets (from the second part of Cantor *1895a*), which involves the non-distinction between cardinals and finite ordinals.

In *1901a* a theory of progressions (the ordinals) is already offered, but although it is not very clear (like the one for cardinals), it provides the essentials. The theory presents ordinals as 'classes of series', i.e. logical structures. Russell begins by defining ω (the simplest one), but in a really obscure way (Kilmister *1984a* describes the definition as 'mysterious', but he does not realize that Russell modified it in the following article). The truth is that with such a definition Russell is depending on Cantor *1883a*, where ω is the first number after all finite integers. (On the other hand, *1902c* already follows Cantor *1895a* —where ω is defined as the order type of the well ordered sets—, so that it offers a new definition and describes the former version as mistaken and rather defining the first transfinite cardinal.) *1901a* continues by defining induction, the ordinals 0 and 1, likeness (still called similarity) between progressions (as the one-one relation whose domain is one of them, its converse domain the other, and such that the predecessors of them correspond to one another), operations in logical terms, inequality, and rationals (ordinals) as operations among integers.

1902c sets the definitive ground in having Cantor *1895a* available. Russell begins by defining the class of relations generating series (Σ) as those contained in diversity, transitive and connected, and he continues with the class of relations generating well ordered series (Ω) (this is the main difference from the former article) as those having a first term and such that any class contained in them has an immediate successor (if any). In the definition it is worth emphasizing the construction '$\pi\text{-}\breve{\pi}$' for the first term, i.e. the only member of the class of predecessors which are not successors (which also gives rise to define the last term and the series without beginning nor end). Afterwards he offers the new definition of the ordinal ω mentioned above: it is the class of well ordered infinite series (i.e. such that $\pi \supset \breve{\pi}$) P such that, if a class (s) contains the first term ($\pi\text{-}\breve{\pi}$) of one of them (p) and the immediate successor ($\breve{\pi}_1$) of all its terms ($\pi \cap s$), then it also contains the whole series. However, Russell offers only the symbolical version (*1902c*, *1.35):

$$\omega = \Omega \, infin \cap P \ni \{s \in Cls \, . \, \pi\text{-}\breve{\pi} \supset s \, . \, \breve{\pi}_1 (\pi \cap s) \supset s \, . \supset . p \supset s\} \; Df.,$$

i.e. ω is defined through the fulfilment of mathematical induction by such series p. The definition from *1901a* is simplified because there the one-one relation of likeness was indicated as similarity between series, whereas here it is already introduced as *likeness* between relations, which will give rise to define the class of relations like one given relation and the class of ordinals.

---

[1] Thus, *likeness* is for relations the same as *similarity* is for classes.

4. The principles of mathematics   171

Likeness between two relations P and P' is defined as the one-one relation S such that its domain is the field of the first one and the second the relative product $\breve{S}\,PS$. Russell again offers only the symbolical version (*1902c, *2.1*):

$$(P)\,L\,(P')\,.\,=\,.\,P, P' \in Rel\,.\,\exists 1 \leftarrow 1 \cap S \ni (\sigma = \pi \cup \breve{\pi}\,.\,P' = \breve{S}\,PS)\ Df.$$

So, like in his whole treatment of relations, the view of them as 'steps' has pre-eminence. The following represents the former concepts (by developing a draft in Couturat *1904a*, II-III, 236):

$$
\begin{array}{ccccccc}
x_1 & P & x_2 & P & x_3 & P & x_4 \ldots\ldots\ldots\ldots\ldots\ldots\ldots\ldots (\pi \cup \breve{\pi}' = \sigma) \\
\downarrow & & \uparrow & & \downarrow & & \\
S & & \breve{S} & & S & & \\
\downarrow & & \uparrow & & \downarrow & & \\
y_1 & P' & y_2 & P' & y_3 & P' & y_4 \ldots\ldots\ldots\ldots\ldots\ldots\ldots\ldots (\pi \cup \breve{\pi}' = \sigma'),
\end{array}
$$

where any step can be expressed in terms of the rest of relations, for example the step $y_2\,P'\,y_3$ would be equivalent to the relative product $\breve{S}\,PS$:

$$(y_2\,P'\,y_3) = (y_2\,\breve{S}\,x_2\,.\,x_2\,P\,x_3\,.\,x_3\,S\,y_3) = (P' = \breve{S}\,PS)$$

Finally, we have the class of relations like one given relation (*2.11) and the class of all ordinal numbers as the class of well ordered similar relations (*2.12):

$$P \in Rel\,.\,\supset\,.\,\lambda P = Rel \cap P' \ni \{(P)\,L\,(P')\};$$
$$N_0 = Cls \cap x \ni \{\exists\Omega \cap P \ni (x = \lambda\,P)\}.$$

The article continues by stating the properties of likeness, defining the operations of well ordered relations and, in general, transcribing the basic theorems of Cantor *1895a* until reaching the 'e' numbers.

In POM we have not many important things added, except the insistence on the philosophical viewpoint, the criticisms of Peano and Dedekind and the relation-numbers.[1] The presentation of

---

[1] Here (like in 4.3.2 above) one notices that the general position of POM seems to return, to some extent, to that of *1901a*, perhaps because *1902c* was later to the finishing of POM (see footnote 2, page 163). The quotations at the foot of the page in *1902c* (pp. 229 and 253) seems to have been added in proofs.

the transfinite ordinals (ch. 38) is twofold: firstly that according to Cantor *1883a* in terms of the pre-eminence of ordinals; secondly that according to Cantor *1895a* (from §296 on) in terms of well ordered series. This situation must have caused the doubts as to whether the former or the latter must have the primacy (§§230, 232), which was finally solved in favour of cardinals, by mentioning their greater simplicity in presupposing only one-one relations and not serial ones (although he resorts to the notion of presupposition, a ballast proceeding from the manuscripts that here cannot rest any longer upon implication).

Russell insists that the definition in terms of likeness of relations supposes the 'correct analysis' of ordinals (POM, §231), by resorting to the view that relations are sufficient to constitute order so that once they are given their field are given through them, with no need for resorting to the terms in themselves (see 4.3.4 above). According to that, Russell replaces the supposed common property of the class of like relations by this same class, according to the principle of abstraction (which was not mentioned in *1902c*, at least by its name), arriving thus at the concept of (ordinal) number of a given (finite) serial relation as the class of relations like it (§231). The parallelism with cardinals could not be greater.

The idea of relation-numbers was already present in *1901a* (p. 31) under the form of an arithmetic of relations, doubtlessly inspired in the arithmetic of Cantor's order types. In POM there is a passage (which seems to have been added as a sort of complement to the chapter on correlation of series, and reminds us of the ideas from *1902c*) that clearly anticipate the presentation of PM (POM, §253):

> We may define the *relation-number* of a relation P as the class of all relations that are like P; and we can proceed to a very general subject which may be called relation-arithmetic. Concerning relation-numbers we can prove those of the formal laws of addition and multiplication that hold for transfinite ordinals, and thus obtain an extension of a part of ordinal arithmetic to relations in general.

The criticisms of Peano and Dedekind have a common idea: they both really defined, not numbers, but *progressions*. Peano, on the one hand, because he offers a structure of axioms from which numbers are obtained by abstraction; but for Russell this did not give the true meaning of the three arithmetical indefinables nor permitted it to recognize them (POM, §122). We have already seen how Russell introduced his definition of cardinal number (in the line of Burali-Forti); here he adds only the possibility of a global definition of the class of classes satisfying the five axioms.[1] On the other hand, he also identifies Dedekind's notion of 'singly infinite system' with his own concept of progression (§239), rejecting the equivalence between ordinals and the relations defining them. The arguments are the following: (i) progressions can put non-numerical

---

[1] This had been already made by Burali-Forti. Russell adds only the identification of this class of classes with the first transfinite number: $\aleph_0$ (the same definition of ω offered in his *1901a*) in terms of the logic of relations (POM, §123). On the other hand, this procedure is very similar to the one used in geometry, where several spaces are defined as the class of entities satisfying a set of axioms.

entities in order (which can be accepted); (ii) ordinals have to be intrinsically something in themselves in such a way that they can be seen 'to the mind's eye' (§242). Of course this last condition seems to be only another sign of Russell's obsolete essentialism, which, besides, is hardly compatible with his own assertion (see above) that it is sufficient with relations in themselves to determine their own field, so that it is not necessary to point out the terms involved.

This brings to light a deep problem we have seen in other places, which now comes to be joined to the former doubts concerning the priority of ordinals or cardinals. Russell recognizes that in mathematics only the ordinal properties are used (i.e. the fact that numbers form a progression). However this view seems to be inconsistent with his whole theory concerning mathematical definitions, for he adds that the logical properties are 'wholly irrelevant': 'By the logical properties of numbers, I mean their definition by means of purely logical ideas' (§230). On the other hand, his theory of cardinals stated that only logical definitions give the true mathematical meaning of number, which seems to be outside of the possibilities of 'the mind's eye' (see 5.3 below).

The position of PM (which already appeared in *1908a*) follows the stated ideas. Firstly Russell defines ordinal similarity (PM, II, p. 295; see also *151.1), afterwards relation-numbers as classes of ordinally similar relations (p. 320 and *152.01), well ordered series through the corresponding class (PM, III, *250.02), and finally the class of ordinals as the relation-numbers of well ordered series (*251.01). Here are the four definitions:

$$P \text{ smor } Q \ . \equiv \ . \ (\exists S) \ . \ S \in 1 \to 1 \ . \ C`Q = \breve{D}`S \ . \ P = S|Q|\breve{S} \ ;$$
$$Nr`P \ = \ \overrightarrow{smor}`P \ Df;$$
$$\Omega \ = \ Ser \cap Bord \ Df;$$
$$N_0 \ = \ Nr``\Omega \ Df.$$

'*Bord*' —*bene ordinata*— being the class of well ordered relations (previously defined) and '*Ser*' the notion of series (which we saw above). In this way the parallelism with order types is emphasized, for they would be the relation-numbers of series in general (*Nr `` Ser*). Once more, all these entities are introduced without resorting to the principle of abstraction.

4.3.6. Real numbers

In the unpublished manuscripts there is no theory of irrationals (that from POM1 has not been preserved). However, the general rejection of Dedekind and Cantor allows us to infer that this theory must have other foundations, perhaps after all in relation to segments of rationals. We have the same situation with the publications prior to POM, for they do not directly consider this subject. Only in *1901a*, on considering a theory of segments in relation to compact series (§§5-7), these segments are presented as a generalization of the theory of real numbers and Peano's article on irrationals is mentioned (see 3.2 above).[1]

Russell introduces in POM his theory of reals: 'A real number, so I shall contend, is nothing but a certain class of rational numbers' (§258). With that he rejects the traditional theory (that of Peano) according to which irrationals are limits of series of rationals having no rational limit, and he rests on what constitutes the ground of any constructive definition: that the defined class has all the properties attributed to the entity usually established. Afterwards he completes the theory by presenting classes of rationals as 'segments' (those infinite classes of rationals being lesser than a given one) and concluding: 'My contention is, that a segment of rationals *is* a real number' (§259). The application of the principle of abstraction consists in founding the chosen alternative opposite to Peano's definition by abstraction, according to which reals are limits of segments of rationals whose equality is used to state those numbers (see 3.2 above). The nominal definition eliminates the notion of upper limit by identifying it with that of segment; with that, the second gap in the *Formulaire* disappears towards a unique chain of definitions starting from logical primitive ideas.

Nevertheless, also here there are problems of historical priority. In 3.2.3 I showed how Peano presented Pasch's theory identifying reals with segments of rationals as an improvement of Dedekind's theory. Likewise, I explained how he introduced his own theory of upper limits as an improvement of Pasch's. However, I also said that he admitted the possibility of identifying the upper limit with the corresponding segment, and therefore the construction of an alternative theory. Consequently, Russell POM follows Pasch's idea through Peano's indication, although he does not clearly recognize it: 'This theory is not, so far as I know, explicitly advocated by any other author, though Peano suggests it, and Cantor comes very near it' (§258). It is true that at the end of the chapter Russell adds a note (pp. 274-5) referring to Peano's article on irrationals where he writes that his theory *was contained* there, including the recourse to segments, but he continues without mentioning Pasch. This would be confirmed by a letter to Jourdain, who pointed out Pasch's anticipation, where Russell replied: 'I did not know what you tell me about Pasch. My knowledge of theories of irrationals is chiefly derived, so far as I remember, from Stolz's *Arithmetik*' (March 21, 1910; in Grattan-Guinness *1977a*, 129; see also MPD, 56; Couturat, on the other hand, clearly mentions Pasch's priority in his *1904a*, IV, 667).

However, even Dedekind had written that every cut in the series of rationals remains firmly stated by resorting to *one* of the two classes it yields (Dedekind *1872a*, 15). Therefore, Russell's later criticism against the supposed arbitrariness of Dedekind's method is hardly justified (Carnap started the custom to attribute to Russell a theory 'simplifying' Dedekind's one in his *1931a*, 42-3). Thus, when he points out the lack of proof of the existence of the cuts, or when he says on Dedekind's irrationals that 'they are merely specified, not defined' (POM, §268) (for they

---

[1] However, already in *m1900c*, the English version that served as a ground for *1901a*, reals are stated as segments of rationals: 'The real numbers are to be considered actually as segments of rationals, not as limits of series of rationals. The axiom of abstraction is fully satisfied by these classes, which give all that need be supposed common to two series of rationals having (as is usually said) the same irrational limit' (*m1900c*, 42). This clearly anticipated the position of POM.

presuppose the existence of the limits of the corresponding series), he seems to be simply repeating the old criticisms made to Dedekind from the formalist viewpoint (see my *1987a*, 140 ff). In later works Russell even arrived at describing Dedekind's construction as mere postulation, with the same advantages 'of theft over honest toil' (IMP, 71).

Finally, after criticizing Cantor's theory[1] (see Grattan-Guinness *1976a*, 170-1), Russell concludes: 'an irrational actually *is* a segment of rationals which does not have a limit; while a real number which would be commonly identified with a rational is a segment which does have a rational limit' (POM, §220). However such a conclusion, which is presented as necessary for philosophical rather than mathematical reasons (§261, p. 275), seems to be inconsistent if we remember that, according to Russell's own goals, *mathematical* definitions must have the pre-eminence in arithmetic. It is not worth the effort to explain the theory appearing in PM (*310), and repeated in IMP (ch. VII), because it is essentially the same one adding the corresponding sophistications (especially when Russell himself mentions the exposition of POM as the advised source; PM, II, 316). In any case, we find no traces of the principle of abstraction in the entire development of the subject.

I shall not consider continuity and transfinites. The dependence from Cantor is here so great that PM can be regarded, at this point, as a mere transcription of his theories to logical terms. Besides, I have explained elsewhere Cantor's continuity (see my *1987a*, 166 ff). As for transfinites, the more relevant points have already been considered in dealing with the infinite class (4.3.2), cardinal and ordinal numbers (4.3.1 and 4.3.2), and in evaluating Russell's acceptance of Cantor (4.1.3). (For an analysis of Russell's positive arguments on the infinite, see my *1987a*, 487 ff).

4.4. Geometry

When Russell attended the Paris Congress of 1900 he already possessed a great deal of experience on the philosophy of geometry. His FG (see 1.5 above) was an attempt to reduce all geometry (projective and metric) to a small number of axioms, which were supposed as obvious, a priori,

---

[1] The criticism of Cantor's theory especially affects to one point (POM, §266): he can define real numbers only through his fundamental series, which presuppose the existence of the limit of such series. However, as Grattan-Guinness has pointed out (*1976a*, 170-1), Cantor defined simply real numbers through the series, i.e. by associating them with their limits, but avoiding any (ontological) identification with these limits. It seems, therefore, that Russell followed merely the presupposed pre-eminence of his nominal methods before the 'abstraction' of Cantor and Peano. Grattan-Guinness does not mention the antecedent of Pasch and describes Russell's theory as 'a significant newcomer', but he rejects Russell's criticisms as depending on 'misrepresentations' and finishes by describing the choice between the nominal theory and the one based on abstraction as a matter of taste.

and presupposed in any 'form of externality'. However, this latter notion was very obscure and there were many difficulties to logically articulate the notion of presupposition, and also to overcome the metaphysical problems of a relational theory of space through the introduction of a hypothetical 'matter' (see 1.5 above and my *1990a*). All these problems made this book a methodological attempt with no continuation. In the next articles there were some novelties mainly derived from Whitehead and Pieri. But the geometric applications of Whitehead *1898a*, although showing the power of Grassmann's methods, start from the concept of vector, which seemed to presuppose elementary geometry (point, straight line and distance) so that they seemed to present a difficulty for a reduction to simpler principles (see Couturat *1904a*, VI, 231).

As for Pieri, although Russell was impressed by his *1898a* and his presentation of geometry as a formal calculus, he did not manage to fruitfully assimilate such ideas because of the lack of a powerful logic, which would be later provided by Peano. It is true that Pieri presupposed this logic (in fact he recognized that his essay was the transcription to natural language of an investigation firstly made through Peano's symbolic logic), but he did not provide any exposition of it. Moreover, the geometry of the manuscripts (see 2.3 above) did not solve the problem of whether space must be regarded as a set of relations. On the one hand, there was the problem of falling into Leibniz's relational theory (i.e. of returning to FG); on the other, the reductive effort indicated that relations were unavoidable, but, again, Russell did not have an operative logic of relations available (which was obtained also through Peano's recourses; see 4.2.2 above). By making use of this new logic Russell returned (as we shall see below) to a relational theory of space, which was an unavoidable consequence that required new ideas and generated new problems with regard to Moore's philosophical ground.

Peano's geometry provided Russell with a viewpoint characterized by the reduction to simplest elements, into a framework partially inclined to logicism, especially in presenting the geometric calculus as something close to logic, in introducing motion through correspondence relations (mappings), and in criticizing the notion of dimension (see 3.3 above). Pieri's improvements, on the other hand, showed the idea of order as allowing a further progress on the same line, especially in making a more logicist vision of the constructions possible (although for Pieri order was a notion outside from logic). Even his vision of the straight line and space gave rise to regarding them as (sets of) logical relations (see 3.5 above), which allowed Russell to think of overcoming the third (and last) gap of Peano's *Formulaire* (the two first being cardinals and reals). Finally, the acceptance of Cantor's continuity allowed him to present geometry as an extension of the idea of series and, therefore, as being definable starting from order, which had already served to reduce great parts of mathematics to logic.

The logicist vision of geometry was already present, at least in its more general features, in the articles of 1901 (see 4.1.2 above).[1] *1901d* already shows the connection to arithmetic through the

---

[1] In *m1900c* there are three (later discarded) sections (§§2, 7, 8) dealing with the definition of the concepts of *group*, *distance* and *angle*, in logical terms. This shows that Peano's influence on Russell's geometry started from his presentation of certain transformations (e.g. motion) as correspondence *relations*, since groups are defined by Russell as classes of one-one correspondence relations.

concept of series, which is reduced to logic by means of the algebra of relations (p. 91). This led Russell to a vision of geometry as independent of space (which clearly proceeded from Peano and Pieri), to be strengthened by the fact that (as it is shown by non-Euclidean geometries) geometric reasoning is strictly deductive and therefore independent from visual intuition. However, the idea of regarding the straight line as a mere relation still did not appear.

1901c progresses towards such an idea by interpreting Peano and Pieri in terms of the new logic of relations and by regarding space as a class of entities (points) related in some way. Thus, after introducing transitive and asymmetrical relations between points, Russell writes: 'All the points forming what may be called the extension of a given relation of the class are a series which is endless and compact... Such a series is called a straight line'. However, in the following Russell made use of non-logical concepts and admitted the point as an indefinable (see 4.1.2 above), which is a proof that Russell was intending only to strengthen his old 'absolute' theory of space and time, which was more favourable to Moore's philosophy (as he explicitly recognizes in p. 20). Curiously, this undefinability of points (which was overcome some years later) led Russell again to some extent to the relative theory. In fact, the whole structure of the geometry of POM was already anticipated in *m1900b* (pp. 13-14), where we find the division of geometry into: (i) *projective*, based on the straight line as a symmetrical relation whose points have no possible order and where there is no 'between' relation; (ii) *descriptive*, admitting two senses for the straight line, the 'between' relation, and the segments; (iii) *metric*, characterized by the notion of motion (which was introduced by Peano as a transformation leaving invariable the property of distance).

POM starts already from a logicist view of geometry as a branch of pure mathematics being rigorously deductive and independent from its possible applications to actual space. Therefore, geometry is indifferent concerning the problem of its particular premises as well as the existence of the entities defined. By adding the reference to order, Russell proposes this definition: '*Geometry is the study of series of two or more dimensions*' (§352).[1] The definition of dimensions takes place also in logical terms: a series of one dimension is a series (i.e. a transitive and asymmetrical relation) whose terms are individuals (points). A series of two dimensions is 'the total field of a class of asymmetrical transitive relations forming a simple series', i.e. a relation whose terms are also relations (§354). This view has the advantage that here dimensions, 'like order and continuity, are defined in purely abstract terms, without any reference to actual space' (§356), so it is the logicist guarantee which presents space (like Peano and Pieri) as a true *construction* (although Russell adds that it starts from the logical and real analysis of space). However, as we shall see, it offers the difficulty that it presents space as a mere set of relations, which finally leads to Leibniz and to disregard points, the only 'absolute' ground for the system.

Starting from that, Russell considers the three mentioned geometries. I shall emphasize only two ideas: the respective definitions of 'straight line' (or the equivalent entity) and 'space', and I shall devote the rest of the section to the philosophical problems so posed.

---

[1] POM, §352. The fact that in this way complex numbers (bidimensional series for Russell) are embraced is not a sign that they depended on real space.

Projective geometry begins with points, every two of them determining a class (usually a straight line), which is reinterpreted by Russell through a relation R, between the points a and b, exhibiting the symmetrical, transitive and aliorrelative (so that $a\overline{R}a$) properties (POM, §363). Afterwards he defines the plane through these only two indefinables and, through the axioms from Pieri *1898a* (including some additions), the rest of projective concepts. However, the two indefinables and all the axioms (i.e. *Pi* and *Pp*) *disappear* by means of the definition of 'projective space' (POM, §413):

> A projective space of three dimensions is any class of entities such that there are at least two members of the class; between any two distinct members there is one and only one symmetrical aliorrelative, which is connected, and is transitive so far its being an aliorrelative will permit, and has further properties...

Then we have a list of the fundamental axioms, which are *incorporated* to the same definition, which obviously is constructed in purely logical terms in starting simply from two terms of a *class* and a *relation* (determined by its logical properties) between them. The first axiom is the existence of the straight line (a term foreign to the field of the former definition: the points); the second one the existence of a third term outside the class, etc., etc. The purely structural character of the definition remains emphasized (like Pieri did on defining space): 'Whatever class of entities fulfils this definition is a projective space'. Consequently, the whole essence of a projective space lies on being a class of relations with certain properties whose terms are indifferent, exactly in the same way as in the definition of order, where we saw how, given the relation, its field was stated, which allowed the replacement of the class of terms (e.g. the series) for the relation itself (which generated the series).

Russell's *descriptive* geometry simplifies (following Vailati) Peano's (which required two indefinables: *point* and *between*). He starts from a class K of transitive and asymmetrical relations whose field *is* the *straight line* that is defined as the class *point* (POM, §376). The enumeration of the properties (axioms) defining the members of this class state the connections between the points and such relations (the straight lines) through only logical concepts: terms, formal properties of relations, domains, membership and implication. Therefore this geometry starts from only one indefinable: the class K of relations. However, on offering (a draft of) the definition of descriptive space this only indefinable also disappears: all its axioms can be formulated in terms of the logic of relations (§378). Like before, a class C of relations (such that $K \in C$) is defined satisfying those axioms. The method is, then, the same one: 'The axioms then become parts of a definition, and we have neither indefinables nor axioms', with which the terms disappear (as such terms) since any term of the field of relations K is a descriptive point. Starting from that, Russell introduces the rest of concepts.

Metric geometry is constructed through the same method; the only difference is that Russell starts here from some of the more general former types and then he adds several specifications (§388). Russell offers several 'logical' definitions of straight line (§395) (one of them proceeding from Pieri) and defines *distance* in terms of the 'magnitude of divisibility' (§396), an ordinal logically indefinable notion, despite the fact that it can be expressed in terms of the logic of relations (so that it is claimed as foreign to pure mathematics, unlike in *1901c*; §411). The definition of metric space is not offered (not even a draft), but it is claimed as following the same model (§413).

All this application of the logicist idea (which I have explained only through his more relevant features for my goals) is a true set of the ambiguities, inconsistencies and difficulties of Russell's method. This same situation permits us also to emphasize the multiple links of his philosophy at this stage. In particular it shows the reciprocal relation between the problem of philosophical and mathematical definitions, the attack against definitions 'by postulates' of Peano (and Dedekind), the underlying ambiguity between axioms and definitions, and the indecision between absolute and relative theories (i.e. the doubts between points and relations as the ground for the system); all that in the general framework of a certain non-distinction between the features proceeding from the unpublished manuscripts and the new stage. Let us briefly consider these points.

First of all we have the apparent pre-eminence of relations over terms: 'When a set of terms are to be regarded as the field of a class of relations, it is convenient to drop the terms and mention only the class of relations, since the latter involve the former, but not the former the latter' (POM, §413). This supposes that in geometry the important thing is only relations, a philosophical view Russell was defending for many years. However, Russell insists also on the mathematical viewpoint, which is indifferent to the kind of considered entities if they preserve the defined relations (§387). All that suppose a certain return to FG (where finally Russell introduced 'material' points to provide the terms for the specified relations) and therefore to the relative theory (which was confirmed afterwards). Furthermore, some other connections with the Kantian position of FG appear, as for instance the need for preserving the deductive relations between the adopted axioms and the observable facts (§421).

Points, regarded as absolute entities, were of course rejectable. On the one hand they were indefinable because they could be (still) not constructed through purely logical methods (Radner *1975a* described the situation as a logicist 'trick'); on the other, the return to the relative theory seemed unavoidable starting from the strong attacks against the concepts of point and space implicit in the surprising correspondences found by Cantor and Peano between lines, surfaces and figures (see 3.3 above). If dimensions were somewhat dubious, then the only differences between a segment, a square and a cube cannot be expressed in terms of classes of points since when we regard them in this way they are equivalent to each other (this situation seems to have led Russell to define dimensions in terms of series). Therefore only the order among these points must be considered, and as we already know, order is nothing but a set of relations.[1]

---

[1] Couturat *1904a* (VI, 813) explained it in this way: 'ce qui constitue proprement et essentiellement les continus à plusieurs dimensions (comme le continu linéaire), ce n'est pas un ensemble de points, mais un ensemble de relations. Ce fait a une portée philosophique manifeste; il signifie, en somme, que l'espace n'est pas une simple "multiplicité", mais bien une multiplicité *ordonnée*; et il justifie la conception de *Leibniz*, qui voyait dans l'espace, avant tout, un *ordre*'.

It was a new argument against points, which Russell clearly expresses when he defends the possibility of eliminating terms in favour of relations. This perfectly coincides with the method (which was already considered with regard to the criticism of Peano's definitions by postulates) of transforming the axioms into parts of a definition. Such a procedure would achieve the transformation of those definitions into *nominal* definitions and, at the same time, the elimination of certain indefinables through the mere recourse of 'defining' them as that which fulfils certain properties. After all it is the same suggestion already made by Pieri, Burali-Forti, Hilbert and Poincaré (see 3.5 above and my *1987a*, 81 ff).

Thus, Russell wrote that a projective space is any collection of entities satisfying those axioms (POM, §§362, 413), with which he made use of a recourse that he criticized in Peano when this author regarded his arithmetical indefinables as 'defined' through the arithmetical axioms. We cannot argue that here there is a greater intuitive ground in starting from a geometry considered as formal calculus where there are so many possibilities as sets of axioms we want to introduce; or, vice versa, that in this way 'what is understood' by geometry resulted clearly *designated* by the definition, since at the same time Russell wrote that Peano's axioms did not define the numbers of everyday life. Moreover, Russell does not provide a logical definition of space which, at the same time, can be regarded as a bridge with 'everyday' geometry, which he actually did for the logical definition of cardinals. Of course the two definitions (number as being an indefinable appearing in several axioms, and space as being a set of axioms) belong to very different levels of abstraction, but I think the philosophical comparison is useful to understand the evolution of Russell's method.

Perhaps this problem can be expressed by saying that a certain lack of appreciation of the axiomatic method underlies Russell's efforts. In fact he could have regarded Peano's axioms as defining an 'arithmetical space' (a structure, rather than a set of indefinables) where the considered entities were *any* class of objects satisfying the axioms. Consequently, the relevant thing to know is whether a set of axioms *determine* a real object in a unique (and formal) way. I think this is the main ambiguity underlying Russell's difficulties in constructing and philosophically interpreting abstract structures, and the only trait that can explain to some extent the tension between formal and intuitive points of view in the whole POM.

This is the reason why Russell could not admit the complete arbitrariness of definitions, in the same way that he could not admit all the possibilities of interdefinability among logical constants (see 4.2). Any complex definition, we are told, is arbitrary only 'within certain limits' (POM, §413), doubtlessly due to the difficulty of admitting a sharp separation between mathematical and philosophical definitions. In fact Russell even says that the identification between axioms and definitions is 'the true way, philosophically speaking, to define mathematical notions' (§378). The need for holding the absolute character of constructions is also underlying all that; that is why Russell describes space as a collection of points instead of as a unity indefinable as a class (§422). Thus he tries to fill the indeterminateness of definitions by means of intuition. This would explain why Russell preserved the third part of the paper he submitted to the Paris Congress (*1900b*) as chapter 51 of POM, which was devoted to maintain the *absolute existence* of points through a strong attack against the relative theory.

In the last analysis we return again to the starting-point: the general ambiguity and hesitation between terms and relations. If points are to be maintained, then the straight line must depend upon them, with which it ceases to be a mere 'relation' *1903b*, 190):

> The straight line, it is true, is generated by a relation, but this relation holds, for a given straight line, between only *some* points and some others... Thus the generating relation of a straight line picks out some points of space as inherently peculiar, so that the straight line, if taken as fundamental, is fatal to thorough-going relativity.

That is to say, only if points have the ontological pre-eminence over the relation can the straight line avoid the relativity of the only formal properties of the relation. Besides, every straight line individuates certain peculiar points, which seems to be destined to avoid the interchangeability involved in the mere formal interpretation of the corresponding definition. However, if the relation has to be something more than these formal properties, then the entire logicist idea vanishes. Russell points out this idea again and again: 'by the *type* of a relation I mean its purely logical properties, such as are denoted by the words one-one, transitive, symmetrical, and so on' (POM, §412; also in §379). He needs this formal character to be able of 'dissolving' relations into logic; at the same time, *he needs the opposite*: that the terms serving as a ground to these relations provide them with their true reality, on pain of falling in a desperate relativity and a complete loss of intuitiveness. This situation could explain passages like this one: 'The straight line may be regarded either as the class of points forming the field of a relation R, or as this relation itself' (§379), with which one cannot avoid thinking that Russell was explicitly resting on this philosophical ambiguity. Of course the rejection of the intensional viewpoint (so dear to Moore's philosophy) avoided this situation (although Russell maintained the old notation).

The geometry of POM was the last one constructed by Russell: the fourth volume of PM was never written, so that the only things we can know of that project remained in Whitehead's later work (although Whitehead ordered the destruction of his manuscripts; see Schilpp *1941a*, 749. See also Harrell *1988a* to know about some extant materials in the correspondence with Russell) and in Russell's constructions inspired on it (see my *1987a*, chapters 14 and 15).

4.5. What Russell learned from Peano

POM did not finish with geometry. The logicist dream intended also to embrace matter (perhaps by preserving the old Hegelian inspiration of the 'transition' to matter) through an attempt to apply the new logic to the same thing which concluded the whole analysis in former writings: the logical foundation of 'rational' dynamics. However neither the definition of *matter* in terms of series (space and time) and relations between them (§441), nor the definition of motion as a correlation among places and times (§446; already present in *1901c*, 305 f), which meet in the definition of a 'dynamical world' (ch. 56), add anything interesting from our point of view. The attempt to eliminate the notion of *cause* through the reduction to a mathematical relation between different configurations of a system (§460) is more interesting, but after the later influence of Whitehead (from 1913 on) the way of Russell's constructions would be very different.

In any case, the main philosophical and methodological result of POM was to have transformed Peano's *Formulaire* into *a unique chain of definitions*, i.e. to have overcome the three 'gaps' requiring the introduction of new indefinables (cardinals, reals and geometrical indefinables). Thus, in achieving the complete logicist construction, the own Peanian project was led to its final consequences (in spite of the break with Peano's vision of intuition and ordinary language). In this way also Russell achieved his old project to rigorously founding mathematics (although when POM was published he was already fighting with the paradoxes). Here we can remember the strong impression he received before the first contact with Peano's *Formulaire* (the main source of the achievement) two weeks after the Congress: 'I am learning Peano's system, which is splendid —the best thing that has been done for a very long time' (letter to Moore of August 21, 1900).

Only by having compared in detail the unpublished attempts previous to the Congress of 1900 with the whole POM it is possible to firmly establish what Russell really learned from Peano.[1] We can summarize the heritage in the following points:[2]
- a conception of the logical calculus which made it possible, through the distinction between membership and inclusion, to overcome Moore's logic, restricted to the whole-part analysis;
- a vision of the logical hypothesis which allowed Russell to include relations as explicit entities in the calculus;
- a set of techniques which made it possible the intensional, precise and rigorous expression of Cantor's infinite, which gave rise to the possibility of offering definitions applicable to numbers in a general way (i.e. previously to the distinction finite-infinite);
- a reduction of logic (and parts of mathematics) to a small set of indefinables and indemonstrables, which allowed Russell to extend the same reduction to parts still not included;
- a theory of implication making the emphasis of propositional logic possible, as well as the identification between implication and presupposition;
- a vision of arithmetic that, although admitting some indefinables, it also offered the way to reduce them into purely logical terms (the 'logicist arithmetic' from Peano);

---

[1] Of course here I am thinking (like with the title of the section) of Kennedy *1973a*, which was only a first step. As he did not work with the unpublished manuscripts (excepting some letters) he did not realize the dependence of the definition of number nor, vice versa, the important precedents contained in the manuscripts previous to 1900.

[2] MPD devotes chapter 6 to Peano, but there Russell recognizes only two main points: the relation of membership and the distinction between the unit class and its only member.

- a set of techniques to eliminate definitions by abstraction (and by postulates) in favour of nominal definitions (Pieri and Burali-Forti);
- a conception of numbers wide enough to suggest new ways of reduction, like the cases of cardinals, ordinals and reals;
- a symbolical notation whose flexibility permitted the transcription of difficult mathematical concepts (as those by Cantor), including many particular examples;
- a set of very powerful logical recourses, including existential and universal quantification, formal implication and implicitly propositional functions, which made the handling of mathematical generality possible;
- some particular notations suggesting new and powerful theories, as for instance the *iota* symbol, which, by isolating the properties of existence and uniqueness, gave rise to Russell's theory of descriptions (including the recourses to eliminate the symbol itself; see my *1990i*);
- a logicist conception of geometry assimilating it to a formal calculus and offering the needed techniques to introduce order and relations in general until eliminating any non-logical indefinables (Pieri);
- finally, a unique chain of definitions, from logic to arithmetic, analysis and geometry. This chain presented some gaps, but it suggested the way to overcome them and the required means to do it.

As a whole, all that constituted a set of techniques of constructing definitions furnishing all the requirements of constructing a logical foundation of mathematics, and at the same time, of *technically* overcoming Moore's vision of the definition as a 'philosophical analysis', while this same vision was maintained as an incentive to preserve the contact with the ultimate philosophical motivations that Russell's method required to continue to be interesting and avoid mere formalism and a complete loss of intuitiveness. It is true that the admission of many constructions required a certain break with intuition and ordinary language, but at the same time it opened the method to the later regarded explicit 'logical constructions'. This, again, hid the old essentialism under new forms, but constituted an unavoidable result (see 5.4 below and my *1987a*, ch. 14).

Nevertheless, we cannot say that Russell inherited a whole philosophical view (as seems to be maintained in Grattan-Guinness *1980c*, 87). This whole view was already constructed through two main traits: to regard definition as the genuine philosophical method, and to try to reduce mathematics (and any other science or concept) to a small set of ontologically simple terms immediately known by intuition. Only in this way is it possible to understand Russell's criticisms of Peano, as for instance the criticism against the extensional view of relations; the accusation that Peano was depending on the subject-predicate pattern; or the argument that he did not distinguish between class and class-concept. I think most of these criticisms supposed some misunderstanding of Peano as a consequence of the philosophical need to preserve Moore's philosophy. In fact almost all of them were forgotten when the new logic obtained the complete pre-eminence (excepting a somewhat vague defence of pluralism and external relations).

The tension between the influences of Moore and Peano comes to light when Russell does not decide between intuition and formalism, as well as when he is forced to go away from the relational theory of judgment to be able to make a full use of the logical techniques, which required to some extent a return to the predicative pattern (implicit especially in propositional functions). Paradoxically, the parallel accusation against Peano turned against Russell himself in so far as the principle of abstraction also represented a return to the subject-predicate pattern through membership. That is why in PM the surrendering to Peano is more complete, once Russell abandoned all attempt to maintain intensionality, the defence of the class-concepts and the principle of abstraction (see 5.2 below for a philosophical discussion of the involved problems).

5. Philosophical and methodological problems

In this final chapter I shall try to reach two goals. The first one will be to consider the philosophical and methodological problems which have appeared in the former chapter, i.e. the logicist reduction, the principle of abstraction and the vision of definitions as constructive devices. The second will be to hold the view that the techniques to construct definitions employed in POM constitute the essence of Russell's method, although of course only a study of the later particular applications of this constructive method can provide the complete evidence (see my *1987a*, chapters 12-15).

5.1. Origin and evolution of Russell's logicism

Russell's logicist antecedents were already present before the 1900 Paris Congress. Dedekind and Cantor had affirmed logical rigour in mathematics, as well as certain technical recourses making the reduction already possible of some parts of mathematics to logic (including set theory), but Russell needed some time to completely realize the advantages of this devices with regard to his own project of the manuscripts, perhaps because he did not yet know Frege. As can be inferred from former chapters, Whitehead's ideas, inherited in part from Boole and Peirce, had also suggested a step towards logicism (at least for certain branches of mathematics). In particular because he offered a unique general construction of mathematics *and* logic out of simple elements and operations. Moore's philosophy came to be a good ally for this attempt because it provided an ontology to the indefinable elements which transformed them into *terms* (or concepts), and also a status to those operations (addition of terms), all of which was accepted by Russell in his unpublished attempts. However, this ontology did not solve the problem because it started from an analysis of ordinary language based upon a scheme (inclusion) always looking for the mere division of concepts into parts, together with the need for regarding number as indefinable.

Russell even knew an explicit version of logicism. Thus, he was familiar with Couturat *1900a*, an article that clearly explained Schröder's logicist interpretation of Dedekind's construction of numbers (although Couturat, curiously, rejected this interpretation with the argument that number is indefinable and intuitive). Therefore, when Russell got in touch with Peano he must have not been surprised by the logicist elements present in his writings. However, for him the most important thing in the beginning was the technical achievements, as may be seen by the fact that the logicist claim did not appear in his first writings after the Congress. In the following I shall consider them with the aim of showing how explicit logicism appeared, and how it was later modified.

There is nothing in the added notes to the manuscripts that can be regarded as a logicist thesis, despite that they show the need to include a whole part devoted to Peano's logic as a true starting-point (see 4.1.1 above). *m1900c* contains a logical definition of cardinal number, but as we saw it was constructed by means of a many-one relation which was finally rejected (due to certain problems of individuation) with the consequence of regarding number as, again, indefinable (see 4.3.1). Not even *1901a* (nor his manuscript *m1901?*) contains a clear logicist claim. It is true that it includes a somewhat obscure criticism of definitions of abstraction, but it is not sufficient to construct the logical reduction suggested by the verbal definition in terms of a class of classes (added in proofs, as we saw in 4.3.1).

m1900b emphasizes that symbolic logic is the essence of reasoning, but these ideas were already old. The assertion that Peano had achieved the transformation of logic into the first mathematical subject (as well as the presentation of all demonstrations as mere logical applications) is, of course, closer to logicism, but Russell immediately adds that the special branches of mathematics (e.g. arithmetic) are to be distinguished by their indefinables. At bottom he still regards (in the line of Whitehead) the logical calculus as 'an independent branch of mathematics' (p. 9) and, although it is prior to other branches, it lacks the genuine indefinables. That is why Russell emphasizes the link between number and logical entities only when he explains the 'second' way of Peano, i.e. that one exploring the logical definitions of certain mathematical entities (see 4.1.2). It is true that a certain interest by this possibility appears, but Russell adds nothing clear enough; not even when he describes the definition by abstraction of cardinal number (p. 13), despite that Burali-Forti had already pointed out the possibility of transforming it into a nominal one (see 4.5.2).

I come now to the four articles published in 1901. Two of them (*1901b* and *1901c*) proceed from an earlier stage, and the third one (*1901a*) has been considered in the former paragraph. Finally, the fourth (*1901d*) already offers the first *complete* exposition of logicism.[1] As I considered its content in 4.1.2, I shall insist here only on the fundamentals. First of all, logicism is presented as an *identification* between logic and mathematics (*1901d*, 75). Mathematics, when pure, consists of implications between propositions (p. 76) and it requires only the logical indefinables. On the other hand, logic can be distinguished by the fact that its propositions 'can be put into a form in which they apply to anything whatever' (p. 76), so that its indefinables (the ones stated by Peano) can be reduced to the notions of *class* and *relation* (in the same line of Burali-Forti), although it requires a new logic of relations (p. 79). Analysis, again, can be reduced to continuity and order, and the same can be made with geometry, although through series of more

---

[1] According to a letter to Jourdain of 1910, this article was written in January 1901 (quoted in Grattan-Guinness *1977a*, 133). The independence from Frege remains emphasized through a note added in 1917 (*1901d*, 79), in which Russell says that at that time he still did not know the German author.

than one dimension (p. 92). As we already know, most of these ideas can be found in the writings of Peano, Burali-Forti and Pieri, so that it is not possible to concede originality to this presentation of Russell's logicism, excepting, of course, his particular formulation and the insertion into a whole philosophical programme.

In POM logicism was completely developed[1] from the philosophical viewpoint. In the preface Russell states two goals. The first one is mainly technical: to demonstrate that the concepts of pure mathematics can be defined in terms of fundamental logical concepts and that its propositions can be deduced from fundamental logical premises. The second goal is more philosophical and it consists in the 'explanation' of those indefinable concepts, which must be carried out as an *analysis* from which those concepts must appear as a residue. This second goal is often forgotten, but it is very important because it gives the link with ontology, i.e. the only thing giving a philosophical interest to the reduction. Of course this kind of analysis must be carried out by starting from the *factum* of logic itself (POM, §10), so that the two goals depend on each other: only the possibility of giving an account of the *existing* mathematics will serve as a criterion for evaluating the list of indefinables and indemonstrables that we 'discover' through the analysis. There we then find the beginning of this 'empirical' (i.e. non-metalogical) attitude of which (after Hilbert and Gödel) Russell is usually accused of.

Starting from the attainment of such goals, Russell specifically claims that mathematics is symbolic logic (POM, §4), so that both can be defined as 'the class of propositions containing only variables and logical constants' (§10). On the other hand, the celebrated definition of pure mathematics states that it is 'the class of all propositions of the form "p implies q", where p and q are propositions containing one or more variables, the same in the two propositions, and neither p nor q contains any constants except logical constants' (§1). Thus, the distinction between logic and mathematics is that, while the first one would embrace the premises of mathematics and the propositions referred only to constants (or variables that do not fulfil the previous condition), the second would embrace the consequents of those premises being adapted to the definition (as well as those premises themselves) (§10). Russell speaks about 'logic' in general, but if we accept that in practice he thinks of symbolic logic (i.e. Peano's logic with certain changes), we can add the definition of the latter he offers: 'the study of the various general types of deduction' (§11). The identification is important since in fact general logic provides him only with the philosophical viewpoint, but the 'empirical' content is the important thing.

---

[1] In *m1902?* logicism also appears, in terms of deducibility from the logical laws: 'From the laws of logic all the propositions of formal logic and pure mathematics will be deducible'. Griffin *1980a* has placed this unpublished manuscript in 1904 with the argument that it contains a criticism of Moore's concept of necessity, which was used by Russell in POM. However, the only place in POM where this theme appears (p. 451) is included in chapter 51, which is nothing but a transcription of the final part of the paper submitted to the Paris Congress (*1900b*). Therefore *m1902?* can perfectly be later to this fact (1900) and *earlier* than POM, once this work was finished in 1903. Maybe (as Grattan-Guinness suggests to me —personal communication) this part of *1900b* was already included originally in POM, but I do not know any evidence of that possibility in the extant unpublished manuscripts.

Finally, 'applied' mathematics will consist in replacing (through a new premise) the variable of an hypothesis by a constant, and in affirming of this constant all the consequences drawn before the replacement has been carried out (§9). The final result is a *continuum* going from logic to 'rational dynamics', where there are no 'gaps' (i.e. no new indefinables) and where all propositions are conditions starting from hypotheses stating the membership of the variables of certain classes. The problem of the arbitrariness of the logical indefinables, which come out when interdefinability appears, is not a difficulty to claim for the existence of a unique set of indefinables (§108). (I have considered in my *1990e* whether is it possible to consider the whole system as a 'formal calculus').

PM eludes, in general, philosophical discussion, but it cannot avoid the claiming of *philosophical* logicism[1] in the twofold customary way: the analysis of the logical concepts and the expression of the mathematical propositions (although here a third goal appears: the solution of the paradoxes). Thus, although Russell does not define here logic or mathematics (through a clear — either Fregean or Peanian— strategy), however he clearly points out (although only once, in the introduction) those main goals of the construction of mathematical logic: 'the greatest possible analysis of the ideas with it deals and the processes by which it conducts demonstrations... the perfectly precise expression, in its symbols, of mathematical propositions'.

The only two subsequent changes were due, respectively, to Wittgenstein's and Carnap's influences. The first one appears in IMP (ch. 18) and is developed through the following implicit reasoning: as logic and mathematics are indistinguishable (we cannot draw a dividing line between them) only *one* definition is needed. This definition will depend on the *formal* viewpoint since both are limited to provide general assertions. Therefore the propositions of logic are characterized by their form, i.e. that what remains invariable when every constituent of the proposition is replaced by another. With that, we are told, we only have a sufficient condition, but not a necessary one (since not all the propositions fulfilling it are logical propositions[2]). The necessary condition is that such propositions must be *tautological*, which makes Wittgenstein's influence clearly appear. Russell does not define this property (he says he does not know how to do it), but he presents it as inherited from the old 'analytical' property.

The second change already involves a certain rejection of the referentialist theory of meaning (which began with the theory of descriptions) and consists in a greater difficulty on identifying the 'constituents' of the propositions, especially if we add the problem caused by Sheffer's discovery of the connective for incompatibility. Hence Russell wrote that although the logical content of a proposition is reduced to incompatibility and the truth of a propositional function, we cannot regard them as genuine constituents without resorting to a Platonic point of view. Thus, logical constants are to be regarded as parts of the language and logic as a more linguistic subject. This

---

[1] MPD (p. 57) contains a more direct explanation: 'The primary aim of *Principia Mathematica* was to show that all pure mathematics follows from purely logical premises and uses only concepts definable in logical terms'.

[2] For example, the axiom of infinity can be logically formulated but cannot be stated, in logic, as being true, since it adds a content incompatible with the tautological property (see also Russell *1911e*, 173).

allows him to preserve formal truth as the main feature of logic, but Russell declares himself incapable of explaining what we mean on affirming that a proposition is true by virtue of its form. In this way, the rejection of Pythagoras supposed mainly an abandonment of mathematics as absolute: 'I have come to believe, though very reluctantly, that it consists of tautologies' (MPD, 157). The climax of this evolution took place in a manuscript of 1951 which has been published posthumously (*1951a*, 306):

> Our conclusion is that the propositions of logic and mathematics are purely linguistic, and that they are concerned with syntax. When a proposition '*p*' seems to occur, what really occurs is '"*p*" is true'. All *applications* of mathematics depend upon the principle: '"*p*" is true' implies '*p*'. All the propositions of mathematics and logic are assertions as to the correct use of a certain small number of words. This conclusion, if valid, may be regarded as an epitaph on Pythagoras.

Therefore, the only things preserved are the importance of the (indefinable) form, the non-distinction between logic and mathematics (through the formal viewpoint) and the notion of truth (as the only criterion to complement the former one).

Here I have no room for a detailed discussion of Russell's logicism through the existing literature (my *1990e* tries to provide such a discussion), so that I come to the analysis of the true philosophical foundation of POM: the principle of abstraction, which, as we have seen, was mysteriously left aside after 1903.

5.2. The principle of abstraction

5.2.1. Origin and evolution

We can say that the first source of the principle of abstraction was the attempt to eliminate properties by replacing them by relations. This attempt was brought to light by Moore and already appeared in Russell's unpublished manuscript AMR as an alternative to monism (see 2.6 above). In it a predicate common to two terms is interpreted as a common relation to some other term. In this way the general plan of the later development of the principle appears, as well as the grounds for identifying the third term with a class. Thus the pre-eminence of *numerical diversity* over *identity of content* is obtained (to express it in the language of Moore's early philosophy; see especially Moore *1900a*). This happens in spite of the similar later pre-eminence of the intensional view of relations (which, as we saw, were interpreted as 'external' due to philosophical reasons), probably because of the logical needs of the subject-predicate pattern implicit in propositional functions (see 4.2 above). In the same way, the need to present the terms of the relations as being also relations to other terms (with the risk of adopting the internal interpretation of relations) contributed to regarding pluralism as the only alternative to Bradley's idealism; but always within the realistic framework as the only position compatible with the Platonic foundations of mathematics. Russell's manuscripts FIAM and POM1 followed the same line according to which asymmetrical relations and the relational theory of judgment constitute the only alternative to the subject-predicate pattern.

There is another topic prior to the Congress of 1900 in which the use of the principle appears: the discussion between absolute and relative theories of position and quantity (see 2.3 above). On quantity, the manuscripts of 1898-1900 show this topic already in a similar way to the later POM, namely, eliminating magnitude as a common property of equal quantities and replacing it by a relation (greater-less) and a class of indefinable terms (every particular magnitude). As for position, the theory appears in a full way in *1900b*, where it is understood not as a reciprocal (symmetrical) relation between related terms, but as a third term to which they have an asymmetrical relation. It leads us to a 'theorem' stating independent series to have pre-eminence over series produced by mere correlation. In the improved version (*1901c*) the theorem states: when there is a serial relation where some terms occur in several positions, then these terms do not form an independent series, but are 'obtained by some many-many or many-one relation each of them has to one or more terms of some independent series' (*1901c*, 297). This paper is later than *1900b* and possibly than *m1900c* too; if so, Russell already knew the formal version of the principle (as we shall see below). However the changes of this paper as regards the first version (*1900b*) are limited to some details. The only sign of a technical formulation is this: 'I consider it self-evident that all symmetrical transitive relations are analysable' (*1901c*, 295)

1901b is an odd hybrid read in May 1900, but reflecting something of the new logic in the final version. (Incidentally, it contains the first reference to De Morgan as a predecessor of a similar principle in terms of the logic of relations.) We read that symmetrical and transitive relations can be analyzed into possession of a common property: 'This again, on examination, is found to consist of sameness of relation to the so-called common property' (*1901b*, 32). This is still a preliminary formulation, but presents the relation to the third term (the property) in a compatible way with the membership relation. We must remember that De Morgan followed Boole in interpreting properties in terms of class inclusion. It was Peano who introduced the membership relation, but Russell's stress on asymmetrical and intransitive relations shows his urgent need for such a relation (independently of other purely logical needs). As we shall see, this need contributed to the problems concerning the distinction between different relations, all exhibiting these properties.

The canonical statement of the principle of abstraction and his application to cardinals was given, as we saw, in *1901a*, but the formulation of the first manuscript version (*m1900c*) is somewhat different. This formulation joins several elements previously defined in a single proposition: many-one relations; relative product of a relation and its converse; and symbolical transcription of transitive and symmetrical relations (this later recourse probably taken from Pieri; see 3.5 above). The (demonstrated) proposition is (*m1900c*, *6.1):

5. Philosophical and methodological problems 191

$$S \in Nc \longleftrightarrow 1 \;.\; R = S\breve{S} \;\;.\; \supset \;:\; R^2 \supset R \;.\; R = \breve{R},$$

i.e. if a relation is equal to the relative product of a many-one relation and its converse, then the original relation is symmetrical and transitive. It must have immediately suggested to Russell a connection with the old and not yet named principle which had arisen in the manuscripts prior to the Congress of 1900: the equivalence between a transitive and symmetrical relation of two terms and the sameness of the asymmetrical relation of these terms to a third one (that is to say, like in POM): '$Nc \longleftrightarrow 1$' being the asymmetrical relation and '$S\breve{S}$' representing the properties of identity as an equivalence relation (although as we shall see below, every relative product of a relation and its converse exhibit these properties). The converse proposition to the former one (i.e. the principle itself), was presented as a primitive proposition (*m1900c*, *6.2):

$$R \in rel \;.\; R^2 \supset R \;.\; R = \breve{R} \;.\; \exists R \;\;.\; \supset_R \;.\; \exists Nc \longleftrightarrow 1 \cap S \ni (R = S\breve{S}) \;\; Pp,$$

which stated the *existence* of the above mentioned asymmetrical relation (although this existence followed merely from the existence of the original relation).

The philosophical consequences (which usually appear in later expositions) were not yet drawn. On the other hand, the link with definitions by abstraction appears, and curiously providing the necessary requirement to make these definitions acceptable, but not, as we would suppose, rejecting them (which was the usual later way; see 4.3.1). The argument consists in the assertion that definitions by abstraction *presuppose* the principle: 'This *Pp* states that, in all cases where definition by abstraction is formally allowable, there exists the entity which is defined' (*m1900c*, 7). And this allows one to use the principle as a ground for Peano's method rather than as an alternative to dispense with it. However, there is at least an instance in this manuscript where the principle assumes the character of an ontological simplification, i.e. eliminating entities in the same way as Ockham's razor: when real numbers are referred to as segments (classes) of rationals and they are based on the principle as a foundation (see 4.3.6 above).

The published version of this manuscript on relations (*1901a*) already offers a demonstration of the principle, but the most interesting feature is the appearance of the first clear criticism of definitions by abstraction. It is claimed (like in the first version) that the principle is presupposed in these definitions, but then there is an addition (somewhat obscure): definitions by abstraction 'do not give a single individual but a class', the class of relations *S* (*1901a*, 10). This fact, we are told, produces an ambiguity because there is no guarantee that every relation indicates the same individual. The criticism remains without clarification in the next section (§2) (this is the reason why the *addition* referred to in 4.3.1 above is interesting).

Moreover, nothing is said concerning the philosophical interpretation of the application of the principle, like in the papers of 1902. In *1902b* the principle is only exposed (by Whitehead, on p. 377), but with no application. In *1902c* there is merely a mention of it with regard to ordinals (p. 16), but this mention only points out the replacement of a common property belonging to a class of series by the class itself and the relation of likeness (the corresponding relation, in ordinals, to the similarity for cardinals). In fact the principle is not used in any demonstration: not even the asymmetrical relation to the corresponding class is emphasized. It seems to mean that its use is confined to a vague support to the ontological simplification supposed in saving a primitive entity (a new application of Ockham's razor). In POM we find the same pre-eminent use, where there is no formal demonstration.

I shall not discuss here the several uses of the principle in POM; I have mentioned it with regard to cardinals, ordinals, quantity and position. Now I shall give only the 'official' formulation, and I shall consider its significance below. The exposition in POM is important because it adds the philosophical ground needed to fuse the *three* underlying versions: (i) the older and more philosophical one according to which properties are really other terms of certain relations; (ii) the technical one through which equivalence relations may be analyzed into asymmetrical relations; (iii) the Ockhamian version allowing the elimination of supposed existing entities by replacing them with the corresponding classes.

The first version was applied to struggle with the classical concept of substance, which depended (according to Moore and Russell) on the subject-predicate pattern. Of course this version rested on Moore's relational theory of judgment (invented in essentials by Bradley in his *1883a*, as I show in my *1990g*), and in fact Moore himself gave in his *1900a* a primitive but very interesting exposition of the principle, including an attempt to distinguish between membership and another implicitly contained relation (as we shall see in 5.2.2, this is the essence of the principle). The second version provided the means to emphasize the pre-eminence of asymmetrical relations over symmetrical ones, which have to be regarded 'of the kind of equality' and therefore they are not 'true' relations. This version was used to criticize definitions by abstraction, and it led Russell to the third version, the only one able to constitute the later method of substituting 'constructions for inferences' (the definitive constructive method from 1914 onwards).

The unification was important because it allowed a perfect continuity between the requirements of Moore's pluralism (based on external relations) and the new techniques of framing definitions, provided and suggested by Peano's logic. (At least this must have been the original intention.) The later appearance of a gap requiring Moore's theory of judgment to be abandoned was an undesired (although unavoidable) consequence of the surrendering to the predicative pattern presupposed in propositional functions. However, the unification required too much from the *same* principle. It had to do with philosophical tasks justifying the pre-eminence of the external relations and, at the same time, to play the mathematical role of *demonstrating* that equivalence relations are analysable into asymmetrical relations and indicating the corresponding class (the property).

There is no allusion to these three different meanings in POM, where we are always told about only *one* principle, but several connotations in the various contexts permit us to isolate real differences. The traditional version now attempts the support of the analysis of equivalence relations, whose goal is the sameness of relation to a third term (POM, §157). However,

5. Philosophical and methodological problems   193

immediately it looks for the foundation upon the analysis of the concept of *common property*, which is presented as a successor of Aristotle's subject-predicate logic. In this way the principle must be interpreted as an abandonment of predicates: 'A common property of two terms is any term to which both have one and the same relation' (§157), which leads to the old concept present in the manuscripts of 1898-1900.

By contrast, the technical version, although starting from the same analysis and therefore culminating in the need for a many-one relation, is presented as a derivation of Peano's definitions by abstraction (§210). In such definitions, we are told, an equivalence relation between two classes u and v leads to the admission of an identity between two new entities $\varphi(u) = \varphi(v)$, so that the original relation is analyzed into sameness of relation to the new terms $\varphi(u)$ and $\varphi(v)$. Of course these entities are interpreted as coming from a common property, which is replaced by the asymmetrical relation referred to above. Then the principle of abstraction is presented as the axiom required to make definitions by abstraction valid, because the guaranteed entity (i.e. the asymmetrical relation) actually exists with only the existence of the original relation.

Finally, the third version, exhibiting a great independence from philosophical tradition (excepting from Ockham, of course), already announces the separation between philosophical and mathematical definitions. In this version we have the same beginning (the equivalence relation), but the entity emphasized now is the respective class in terms of *conceptual economy* (POM, §111):

> Whenever Mathematics derives a common property from a reflexive, symmetrical, and transitive relation, all mathematical purposes of the supposed common property are completely served when it is replaced by the class of terms having the given relation to a given term.

The important feature here is not the new asymmetrical relation between every term and the property (or class). It is enough to have the class of terms (united by the equivalence relation) *itself*, because it fulfils all required conditions. We can understand now the reason for using this version when there is a dominance of mathematical purposes (like in the papers of 1902). Then there is no need for resorting to the philosophical and technical reasons, underlying in the two other versions, to justify the method.

PM contains a definitive statement of this last version. There is a technical formulation and a demonstration (in *72, devoted to miscellaneous propositions), but the principle is not actually used and not even mentioned in the rest of the book. The beginning is the proposition *72.64:

$$S^2 = S \ . \ S = \breve{S} \ . \ R = Cnv\,'\,(\overleftarrow{S} \upharpoonright D\text{'}S) \ . \ \supset \ . \ R \in Cls \rightarrow 1 \ . \ S = R|\breve{R},$$

which states that if R is the converse of the relation S (transitive and symmetrical) of the class of relata (limited in its domain), then R is a many-one relation and S is the relative product of R and its converse. From it the principle of abstraction is derived (*72.66):

$$S^2 \subseteq S \ . \ S = \breve{S} \ . \ \equiv \ . \ (\exists R) \ . \ R \in Cls \to 1 \ . \ S = R|\breve{R} \ .$$

However, the supposed philosophical content of the principle is referred to only by mere respect to a venerable (but now useless) tradition. The argument is that in searching for definitions for the several kinds of numbers (PM, I, 442),

> we always have, to begin with, some transitive symmetrical relation which we regard as sameness of number; thus by *72.64, the desired properties of the numbers of the kind in question are secured by taking the number of an object to be the class of objects to which the said object has the transitive symmetrical relation in question.

Nevertheless, the supposed philosophical ground is completely avoided *in fact* because the property is replaced by the class of objects, proposed only on the pragmatic basis that the class in question has the same mathematical properties. Thus the principle is reduced to a mere description of an actually used method without being applied as a real foundation in any formal deduction. That is to say, all definitions and demonstrations in PM would be exactly the same *without* the principle of abstraction. Russell himself partially acknowledges this fact when he added that the principle, 'though not explicitly referred to in the sequel, has a certain intrinsic interest, and generalizes a type of reasoning frequently employed by us' (PM, I, 451). In fact this renunciation of the more genuine foundation of POM (the only link between the new logic and Moore's pluralism through the first version) has been never pointed out. Thus a need arises to *explain* a change which has not been discussed before (including by Russell himself). We can provide such an explanation only if we go deeply into the principle itself and the kind of definition that it is designed to replace (definitions by abstraction).

5.2.2. Assumptions and implications

My main thesis concerning the change is that Russell became aware that the relation really present between the terms of an equivalence relation and the respective class is *membership*. And, as it is asymmetrical but not a many-one relation, the technical interpretation of the principle of abstraction emphasizes only the many-one relation (i.e. a function) between equivalent terms and its equivalence class, and in such a way that if we want to use it we can do so only 'philosophically'. Formally it states only a logical consequence concerning a kind of relations, but does not create *existence*. Thus, only the first and third versions mentioned above are preserved. For although the referred class has all the properties assigned to the terms (Ockham), we replace the common property by a class from which the terms are *members*, and not exactly by a third *term*. However, to replace a property by a class was a technique known already to Boole, and this without the philosophical implications required by Moore's theory of judgment (mainly because of the problems implicit in the attempt to distinguish between a class and a class-concept).

All these elements led to a reconsideration of the *intensional* view about relations and to a victory of the Peanian approach, including the interpretation of equivalence and propositional functions in terms of *membership*. The result of this situation was the formal reconsideration of relations in PM (as mere propositional functions of two variables) and the abandonment of classes as entities. From the *methodological* point of view it means to preserve the principle of abstraction only as vague foundation in terms of conceptual economy. Finally it led Russell to the statement of 'logical constructions', which involved the withdrawal from the old philosophical logicism and the renunciation to find the true logical (and ontological) indefinables. The following is an attempt to justify this thesis.

The early usages of the logic of relations did not involve serious criticism of definitions by abstraction, but the new way to regard relations contained recourses leading it to the idea of the appearance of 'new' entities as a result of some logical operations. Relative product, for example, seems to produce a new term:

$$x\,(R|S)\,z = x\,R\,y\,.\,y\,S\,z.$$

In the case of asymmetrical relations this produced the impression of creating *existence* (for instance, 'x is the grandfather of z' means there *exists* a third term y which is son of x and father of z). In the same way, the theory of absolute position (or quantity) allowed the elimination of terms united by an equivalence relation, replacing them by another *existing* term and the respective many-one relation also *existing*. Incidentally, this result was applied to explain the reflexive property: if we have a symmetrical relation R joining a set of terms, we have as well an asymmetrical relation S joining those terms to a third term. Therefore the proposition xRy will be equivalent to the existence of a new term α such that $xS\alpha$ and $yS\alpha$ (POM, §210). Applying this consequence to the reflexive property we shall have that xRx is equivalent to $xS\alpha$ and $xS\alpha$, i.e. we have obtained a class as the new term. The same result happens with series, for given certain conditions, we can obtain an ordering of the terms in which some of them occur in the same position, i.e. the existence of a relation R. Again, applying the principle of abstraction, there arises the relation S in such a way that if xRy we obtain a new 'entity' t such that xSt and ySt. These entities ('absolute positions' in Russell's language) will form an 'independent' series ruled by the relation $\overset{\cup}{S}\,|R|S$, whereas the original terms will form only a series 'by correlation' (POM, §211).

This existential interpretation of the terms of certain relations was also a consequence of the form assumed by the principle of abstraction, a form producing the appearance of a relation that, being many-one, seemed to require a unique new entity. The later identification of this entity with a class assured the 'existence-theorem' ('i.e. the proof that there are entities of the kind in question'; POM, xix). The problem was that if this entity was seen to be a class, one was tempted to interpret the relation to this class in terms of membership. Thus, the anti-abstractionist power of the principle appeared (partially already in *1901a*) to individuate the domain for a given term (the class of its relata) as the *entity* designed by the definition by abstraction, i.e. a class. However this interpretation was, again, a consequence of the form in which Russell seemed to have understood cardinals in the beginning, i.e. through both domains of a many-one relation (see 4.3.1 above). On the other hand, the *addition* made by Russell to the French version of *1901a*, which interpreted the definition in terms of a class of classes may be seen as being ruled by the scheme of the membership relation: every class similar to a given one is a member, as a *term*, of the class including all these classes (and only them).

In PM the new term produced by the principle of abstraction appears clearly as a class (and of course it is symbolized by a small Greek letter). We can read there (PM, I, 442) that, after stating the equivalence (not the equality, as in earlier versions) between every equivalence relation (now S) and the relative product of a many-one relation (now R) and its converse, there is added:

> whenever the relation S holds between x and y, there is a term α such that $xR\alpha \cdot yR\alpha$, where R is a many-one relation; and *72.64 shows that this term α may be taken to be $\overleftarrow{S}\,'x$, which is equal to $\overleftarrow{S}\,'y$.

(For *72.64 see above.) In this way we see clearly the identification of the new term with the class α, which, again, is equal to the class of relata of x (or y), for

$$\overleftarrow{S}\,'x = \hat{y}\,(x\,S\,y).$$

Thus, the interpretation of R as membership of this class seems possible. We *know* that this interpretation is mistaken because R is a many-one relation (a function), therefore is defined as

$$x\,R\,y \,.\, x\,R\,z \,.\, \supset \,.\, y = z,$$

whereas the relation of membership ($\in$) does not individuate a unique class, but intuitively the interpretation *would seem* possible. Although we shall see later that *both* relations are actually involved, however the criticism that Russell applied *in fact* to definitions by abstraction does not seem to dissolve the intuitive impression referred to. Thus, it is necessary to examine such a criticism.

When Russell included many-one relations in *1901a* he seemed convinced of the absolute theory of position, which was then very important by its usefulness against monism, and this task required such a relation. This theory made the appearance possible of 'separate' terms common to those destined to be eliminated (see Vuillemin *1971a*, 33). In POM Russell summed up all his criticisms of definitions by abstraction in this one: definitions by abstraction are ambiguous; they do not designate uniquely their object. This criticism is really the same as the one appeared in *1901a*, although there the situation was very complicated due to the lack of a clear definition of cardinals and the (dubious) role of the principle of abstraction (see 5.2.1 above). In POM Russell adds a point completely within the usual essentialistic mistrust of non-nominal definitions: the requirement of a unique number to each class ('the' number of the class) (§111). However, when the requirement is applied to definitions by abstraction it is interpreted directly in terms of many-one relations, with the argument that only these relations may translate appropriately the idea of *common property*. Thus we are led to the standard definitions in terms of a class of classes and *only then* the criticism against abstraction is expressed. Consequently, there is nothing strange in regarding abstraction as an ambiguous method.

The particular conditions required by Russell to be fulfilled by every such entity (a number) are: (i) that it is individuated by every class, (ii) in a unique way (every class to only one entity), (iii) in a way that all similar classes point out the same entity. The three conditions form a structure similar to the one constructed to replace Peano's five axioms, with the argument that the entities designated by those axioms constituted only a general progression. Now and then, Russell needs the defined entities to be visible 'at least to the mind's eye' (POM, §242), in a way that we can construct the intuitive representation allowing the coincidence between the axioms and the actual numbers of everyday life (one, two, etc.). That is why he holds there is an infinite number of entities satisfying the required axioms. As a consequence, he uses the customary scheme of many-one relations to emphasize *a priori* the genuine entity (the number) with the power to eliminate similarity among classes. However, these definitions in terms of a class of classes only achieve the inclusion as *members* of the classes similar to one another. We arrive, therefore, at membership.

It is true that Russell's definition of number fulfils the required conditions; but he cannot *also* claim that the definition avoids the *intuitive* ambiguity of definitions by abstraction. Here again the principle of abstraction is used in several ways at the same time: philosophically interpreting properties as terms; technically constructing asymmetrical relations; economically (in the conceptual sense) eliminating similarities among classes in favour of a new entity. However it cannot avoid the fact that the only thing actually made is the construction of a class exhibiting the same mathematical properties as numbers because of the existence of the relation of similarity among its *members* (which, again, are classes); and besides without the intuitive content we must have hoped, i.e. without going outside the predicative pattern, in spite of translating it (according to Boole and Peano) into the scheme employing the notions of class and membership. As Weyl (*1949a*, 12) wrote: 'To say that we know a set means only that we are given a property characteristic of its elements'; therefore, he adds, mathematically creative definitions are limited to constructing a class by means of properties. Thus, there is no objection to the definitions of number as a class of classes, but 'it's an illusion —in which Dedekind, Frege, and Russell indulged for a time, because they apparently conceived of a "set" after all as a collective— to think that thereby a concrete representation of the ideal objects has been achieved' (*ibidem*).

As we shall see later, such a conception was present in the kernel of the problem, but before I shall show how the method of abstraction leads us also to the relation of membership if we consider its implications in ordinary language. I shall start from the famous example of the parallelism relation (so dear to Frege and Peano): 'parallel straight lines have the same direction, therefore the direction of a straight line will be the class of lines parallel to it'. According to the principle of abstraction, a many-one relation R exists between parallel lines (many) and their direction as a property defined (eliminated) in terms of a class (one). In this way a class α appears such that all lines parallel to one another have the same relation to it, i.e. $xR\alpha$. $yR\alpha$. This class will be identifiable with the class of relata of each line according to the equivalence relation S. Thus,

$$\alpha = \overleftarrow{S}\,'x = \overleftarrow{S}\,'y$$

(the class of lines parallel to x and y). Finally, we also have that $R|\breve{R} = S$. However in this way neither the relation of membership (as different from R), nor the elimination of the property and the role of $R|\breve{R}$ appear as sufficiently clarified.

However, the class α is only the class of lines parallel to one another and now this class is replacing the property (the direction) and the relation between every parallel line and the class of parallel lines. (This last relation seems to be membership, although we know that it does not coincide with R.) There appears to be synonymy between the expressions 'to be parallel lines' and 'to have the same direction', but this last expression, again, means only 'to be a member of the class of parallel lines' (all that, of course, starting from a given straight line and a given direction), therefore at the end we have introduced nothing and eliminated nothing. Our only move consists in stating a law forbidding (avoiding) the abbreviation 'same direction', but ontologically we have only introduced a class, the class of all lines parallel to a given line, and the existence of this class is independent of the name which we assign to it. The appearance of the many-one relation R produces the impression of uniqueness of the term pointed out by the relation because every parallel line (to a given parallel line) is functionally related with something unique, but this last thing is only the class itself of parallel lines. Again, we only obtain the transformation of a property into a class; i.e. we come back to Boole.

In the case of cardinal numbers the same thing happens. The only difference is that, starting from elements already being classes, the class obtained is a class of classes. Also here Russell emphasizes uniqueness: 'There is a many-one relation which every class has to its number and to nothing else' (POM, §110), so that he can strengthen his criticism against Peano. The point is important because it shows how the many-one relation occurs now between similar classes and a class of classes being the number of those similar classes. If we would apply the relation of membership in this moment it would occur now between two classes, but not exactly between a term (in the example mentioned above, the line) and a class. Obviously similar classes are being regarded as terms, and not as parts (in this case the relation will be inclusion). Thus we come back again to the same problem: R cannot be membership.

If we choose less common examples, synonymy and membership appear too: 'equal segments have the same length'; 'people born in the same place have the same nationality', etc. Now we see how to define these entities the starting recourse is the scheme 'to have the same x', but it leads us directly to membership of a class: the class of all terms having the property, and this, again, is only referring to the class of those terms that are *equal* with regard to something. Therefore we are resting on two synonymous expressions: 'to be equal concerning x' and 'to have the same x'. The cases of number and direction present a more *reductive* aspect only because they contain expressions where, although the 'same' property is really present, the term 'equal' does not appear, but an adjective setting the bounds of the equality ('similar', 'parallel'). Of course, we can formulate our last examples speaking, say, of 'equilengthened' segments, or something like that, to express 'the same length' without the word 'same'. Finally, it seems we only emphasize a synonymy in terms of membership: 'to be equal segments' is equivalent to 'to have the same length' and also to 'to belong to (to be a member of) the class of equal segments' (always with regard to a given segment), and these equivalences have certainly no informative power.

If we ignore the prohibition of considering membership as a functional (many-one) relation our results will be, of course, uninteresting. We can try to fulfil the requirement to point out only one class, arguing that two elements may be members of several classes by virtue of distinct relations stated among them. Therefore, we can say, the membership relation is 'different' and then 'unique' in each case. We can add, in the same way, that in spite of the fact that one given term can be a member of possibly a lot of classes, however it is a member of each by virtue of a unique reason (or a unique combination of reasons), therefore a 'given' membership would be unique. Nevertheless, all this line of argumentation presupposes confusing membership with the exhibition of a property speaking of qualitatively different 'memberships'. Furthermore, if we make $R = \in$ in the logical expression of the principle of abstraction, we must fulfil the requirement to make $\in | \breve{\in}$ be equal to the equivalence relation S. Applying then the corresponding definition of the relative product we obtain:

$$x (\in | \breve{\in}) z = x \in \alpha . z \in \alpha,$$

but this only states that there is always a class α from which our two terms are members (i.e. they have always some common property). The conclusion is doubtless true but trivial. It is equivalent to say there is always a relation between every two terms, and this relation only may be the universal relation. We have then $\in | \breve{\in} = V$, i.e. one of the propositions demonstrated in PM (*62.24). Our attempt, as we already knew, is hopeless.

We must insist precisely on the distinction between $\in$ and R. Of course Russell clearly rejected the possible identification between them, although this may be only a sign to think about the difficulty of intuitively seeing the distinction itself (Moore also held the distinction in his *1900a*). Thus, stating the principle in one of its three versions, Russell said that, despite that equivalence relations arise only from common properties as other *terms* to which the former have only one and the same relation, however this relation cannot be membership (nor the subject-predicate relation): 'For no subject (in the received view) can have only one predicate, and no individual can belong to only one class' (POM, §157). However, he is not very explicit concerning the characteristics of this other relation and leaves the matter with a misleading remark: 'the relation of the terms to their common property is, in general, different in different cases'. A clearer assertion about the distinction follows the standard definition of number as a class of classes (§111):

> Membership of this class of classes (considered as a predicate) is a common property of all the similar classes and no others; moreover every class of the set of similar classes has to the set a relation which has to nothing else, and which every class has to its own set.

We may distinguish *two* different problems here: (i) the vagueness in the referred distinction, (ii) the consequent difficulties of differentiating the cardinal number of a given class from the class of such numbers. We saw above (in 4.3.1) that Russell does not solve the second problem in a satisfactory way neither in *m1900c* nor in *1901a*. However, in *1902b* he offers the definition (which will be maintained till PM) according to which Nc is the class of classes being the cardinal number of some class. However this expression is only the transcription of Frege's definition, replacing the terminology of properties and extensions of concepts by the more flexible one of classes. Frege gave up classes in his *1884a* (§55), identifying numbers with properties of concepts (§46). From this position, his definition of the number of a class is the following: 'the number which applies to the concept F is the extension of the concept "equinumerous with the concept F"' (§68), and the definition of the class of such numbers: 'the expression "n is a number" is to be synonymous with the expression "there is a concept to which the number n applies"' (§72).

This solves the second problem but the first one remains (to use the now usual set-theoretical terminology): the distinction between the equivalence class as a class to which all equivalent elements belong (are *members* of) and the same class as a member of the respective quotient set, to which the equivalent elements have another completely different relation: the one referred to by Russell as a many-one relation. Imbert (*1969a*, 183) has described the point clearly: 'Frege had no means, neither in ordinary language nor in the conceptual notation, to differentiate the extension of a concept, as an equivalence class, from the corresponding element defined on the quotient set'. We have seen this assertion can be partially applied to Russell as well, in spite of his attempts to clarify the distinction. Through set-theoretical terminology we can explain now the distinction as follows. In the case of numbers, the starting set will consist in classes in general; the equivalence relation will be the similarity relation; the equivalence class will be the number of a class (i.e. the class of similar classes); and the quotient set, the class of cardinal numbers. In general, the distinction between the two relations would be the following: the relation between the elements of the starting set and its equivalence class is membership; and the relations between every element of this set and each of these classes in such a way that it ascribes to each element one and only one class is a 'surjective application' (see Queysanne *1964a*, 48). Now we can understand Russell's insistence on uniqueness and the role of membership.

I come now to consider the concept of relative product ($R\breve{|R}$) characterizing the 'analysis' that the principle of abstraction carries out (we are told) with equivalence relations. We need only to understand exactly what the relative product of a relation and its converse *adds* to the principle of abstraction. The most interesting thing here is to know its properties as a relation in order to evaluate its true role. From the beginning we see this relation as symmetrical and transitive (therefore reflexive) in the sense of the logical expression of the principle that is equal to the relation S, and that this relation exhibits such properties. We may, besides, verify these properties in PM. By *34.7 it is symmetrical

$$Cnv\,'\,(S|\breve{S}) = S|\breve{S},$$

and by *34.8 it is transitive (if symmetrical), for

$$R = \breve{R} \,.\, R^2 \subseteq R \,.\, \supset \,.\, R = R^2 = R|\breve{R}.$$

This is demonstrated starting from *34.28:

$$R = R' \,.\, \supset \,.\, P|R = P|R',$$

from which we obtain

$$R = \breve{R} \,.\, \supset \,.\, R^2 = R|\breve{R},$$

i.e. $R|\breve{R}$ is always an equivalence relation. Therefore its role in the principle of abstraction adds nothing to our former analysis from the philosophical viewpoint. In fact, this connection with equality and identity was noticed by Russell already in 1901 in the case of the many-one relations (*1901a*, *5.6):

$$R \in 1 \to Nc \,.\, \supset \,.\, R\breve{R} \supset 1'$$

(1' being the notation for the relation of *identity*), although he explicitly denies that we may have also $R\breve{R} = 1$, with the argument that the domain of $R\breve{R}$ (the same as R) is only, in general, a part of the domain of identity (*1901a*, 9). Likewise, in PM (*71.19):

$$R \in 1 \to Cls \,.\, \equiv \,.\, R|\breve{R} = I \restriction D'R,$$

so that in this kind of relation the members of the class $D'R$ are identical to one another, for by *35.02 we have that

$$I \restriction D'R = \widehat{xy}\,(x = y \,.\, y \in D'R).$$

In such a case $R|\breve{R}$ is the relation of identity referred to the members of the domain of R. (According to the usual intuitive example by Russell, when we have a relation $1 \to Cls$, 'we go by R and back by $\breve{R}$ ' to the starting-point; PM, I, 427-8.) In any case, by *34.702 we know the field of $R|\breve{R}$ to be equal to the domain of R, which allows us to apply identity to the complete field of $R|\breve{R}$. Again, the details seem to add nothing important.

Consequently, we come back to many-one relations. The *uniqueness* provided by these relations points out a supposed entity (the 'third' term) with a kind of ontologically unanalysable existence. Therefore this class was not to be identifiable with the equivalence class or the respective predicate. In the first case we would arrive at a mere 'construction' (Ockham); in the second we would follow Peano's pattern according to which every proposition may be reduced to the standard form $x \in \alpha$ (POM, §33), and this pattern contains a seed of extensionality leading it to the interpretation of propositional functions in terms of membership. (For example 'Socrates is a man' and the corresponding propositional function 'x is a man' have the same pattern, φx, which may be seen as reducible to the membership of x of a class; for details see POM, §§20-2.) Thus, the risk of eliminating the explicit character of relations appears, until presenting them as propositional functions of two variables. (This was the way chosen in PM, really closer to the ordered pair reduction, in spite of preserving the intensional *notation*.)

Therefore, the Platonic role of the separate and independent entities (mentioned above) is revealed. However, once the functional relation is identified, the only guaranteed thing is a correspondence, and by no means the existence of supposed entities not being the equivalence classes themselves. If we add the paradoxes involved in the admission of certain classes of classes, we claim after all the uselessness of the principle of abstraction. This uselessness is also remarkable for ontology, since the only guarantee of equivalence classes is membership itself, and we arrive at membership only by means of a predicate. The final result is the incapacity to create a method of overcoming definitions by abstraction.

Our general conclusion agrees in the essentials with that of Vuillemin (*1971a*, 34 ff), but a curious feature is that in his earlier work on Russell, Vuillemin did not seem aware of the main problem concerning the principle of abstraction, to which he presented only unimportant objections (*1968a*, 188 ff). In his later *1972a* and *1975a* (both concerning precisely this point) he criticized only the Platonic character of the 'separate' entities required by the philosophical version of the principle of abstraction. However, not even in his *1971a* does Vuillemin see the main point: the multiple versions hidden under the 'same' principle made it impossible to discover that the distinction between the two relations involved has as a consequence that the principle itself guarantees *nothing*, excepting the corresponding class, the respective membership by means of a predicate, and the synonymy described above. However, he seems to express a partially similar idea to ours when he writes: 'To secure the uniqueness of the principle of abstraction it is a sufficient condition to put, as an attribute, an attribute defining the membership of an equivalence class' (*1971a*, 38).

To sum up: Russell's principle of abstraction adds nothing to (i) the mere decision to replace a term by a synonymous one in our language, or (ii) the mere 'economical' recourse to technically replace a set of equivalent terms (or classes) by the corresponding equivalence class. In both cases the operation being ruled by the membership relation, that is the property or predicate. This must have been the reason to explain the absence (under the apparent use) of the principle in the papers of 1902 and in PM. On the other hand, the continuous use of the principle in POM was motived by the attempt to give a ground, philosophical and technical at the same time, to the constructive definitions given there for the first time. (Probably doubting on the true implications of the new method and without deciding to regard the definitions as being 'philosophical' or merely 'mathematical'.) With that, Russell was convinced he was making a contribution to Moore's view according to which relations are the ultimate ingredients of the world. This contribution had to be encouraged for the fact that the principle supposed a demonstration of the existence of asymmetrical relations different of mere membership. In fact, Russell's appendix about Frege in POM shows the first signs of a decision to regard relations in an extensional way (although preserving the handy intensional notations).

The final destiny of the principle of abstraction possibly was the (secret) admission by Russell to have been pursuing an illusion, and therefore the admission that to replace a property by another term and a certain relation was hopeless, in spite of its need in Moore's theory of judgment. All this occurred in the paradoxical frame of POM, inside a general scheme completely intensional ruled by propositional functions (regarded as properties) and by another illusion (which arose in Pieri and Burali-Forti) consisting in the belief that nominal definitions (i) give *true concepts* and (ii) designate uniquely, whereas the rest of definitions would give only mere *intuitions* (in Burali-Forti's sense according to which intuitions are neither precise nor rigorous). On the contrary, nominal definitions seem to be the surrendering to Aristotle's subject-predicate pattern (see Geymonat *1955a*, 60), and the elimination of a frame of relations perhaps more coincident with the relational theory of judgment received from Moore (although without its Platonic charm).

The kernel of all these difficulties and tensions seems to lie in the ambiguity of Russell's ontology on the links between terms, classes and relations, in a similar way as Peano after his discoveries concerning the interdefinability of the logical primitive ideas. On the one hand, the intensional view was fundamental to reach the construction (and operations) of infinite *wholes* by means of the single recourse to 'the class of terms having some given relation to some given term' (POM, §330). However, this recourse led Russell directly to: (i) the distinction between a class and a class-concept; (ii) the threat to transform membership into inclusion; (iii) the preservation of common properties. Russell insisted that the thing defined by his definition of number is a class and not a class-concept, arguing that the predicate 'similar to *n*' is harmless and that only inside a class it may serve to constitute terms (the similar classes) (§111). However, he could not avoid the extensionality without the defining predicate, which is necessarily a concept and its only relation with all constituent classes is to *define* a class to which those classes belong as members, i.e. to give a property. Of course, the principle was not needed for this mission. Furthermore, the distinction between the class and the class-concept had as well another unpleasant consequence with regard to Peano's logic: it made the admission of the null class difficult (Russell wanted to admit only the 'null class-concept') and it forced him to identity the class having only a term with that one term (POM, §§69, 73). Such consequences, though harmless in practice, play a certain role in the path towards extensionality of PM.

On the other hand, the influence of Moore was not enough to eliminate the recourse to properties, not even if we remember that one of his main doctrines was exactly to eliminate them in an explicit way through the mentioned relation to other terms (§216). If Russell had applied this method rigorously, propositional functions (φx) also had to be reduced to the relational pattern making $\varphi x = \varphi R x$. (This was approximately the view from the manuscripts of 1898-1900, where the relation between subject and predicate was regarded as a true relation.) However, it was impossible whereas Russell, for other reasons, held the (mainly mathematical) need to distinguish between *things* and *concepts*, i.e. between subjects and predicates. Then, the new logic of relations threatened its status itself, allowing constructions such as

$$x R y \: . \supset . \: x \in \rho \: . \: y \in \breve{\rho},$$

which seemed to eliminate relations replacing them by membership of certain classes (or domains) (POM, §198). In this way, the famous objection by Bradley (*1883a*, ch. 3) according to which relations always involve other relations between themselves and the related terms, was partially recognized, and this fact directly led it to the predicative pattern, so coinciding with the difficulties of Russell's geometry of POM, where there was a great tension between the priority of an abstract structure over its field, or vice versa (see 4.4 above).

Russell seems to admit the unavoidable place of membership as the ultimate essence of the principle of abstraction from PM on, where he mentions the principle always in these terms. Moreover, the principle appears in this period always only as unimportant. Not even in a work such as IMP is the principle of abstraction mentioned (though, curiously, we find in its ch. 5 a method of analyzing asymmetrical relations in terms of membership). In fact, the belief that Russell always maintained the principle (see for instance Vuillemin *1971a*, 33) belongs also to the received view. In fact the principle only appears (after PM) twice: in OKEW and *1924a*, works where Ockham's razor (i.e. the *third* version of the principle as described above) became methodologically preeminent. In the first case we read that, when a group of objects shows the kind of similarity usually attributed to a common property (OKEW, 42; my emphasis):

> the principle in question shows that *membership* of the group will serve all the purposes of the supposed common quality, and that therefore... the group or class of similar objects may be used to replace the common quality, which need not to be assumed to exist.

In the second case the description of the principle is identical. Starting from the same equivalence relation, the supposed common quality (*not really rejected*) is to be replaced by the respective class: 'all the formal purposes of a common quality can be served by *membership* of the group of terms having the said relation to a given term' (*1924a*, 327, my emphasis).

As I proposed at the beginning of this section, only this certainty about the actual essence of the principle made the admission of 'logical constructions' possible, i.e. the abandonment of the original logicist project based on the strict logical (and ontological) undefinability. Thus, the epistemological constructions received a new pre-eminence and the method was transformed into a more pragmatic instrument. In fact, all this was already possible in POM, but it was also necessary to abandon the hope in preserving the relational theory of judgment received from Moore.

5.3. The constructive definition

In this section I come to the true essence of Russell's method: definition. The adjective 'constructive' refers to the simple fact that through his usual nominal definitions Russell explicitly *constructs* the relevant concept out of materials belonging to lower ontological levels. According to their fundamental characteristics, constructive definitions: (i) *analyze* the concept into 'simple' components; (ii) presuppose a *reduction* to these components; (iii) try to use *ordinary language* as a guide; (iv) obtain *precise* meanings; (v) cause a certain *loss of intuitiveness*. At later stages, the need for including the elimination of the defined concepts was admitted due to paradoxes (see my *1989a*); as well as the break with the whole-part scheme starting from the influence of Whitehead (see my *1987a*, ch. 14). In the following I shall consider firstly the requirement of *nominal* definitions, secondly the distinction between mathematical and philosophical definitions, thirdly the problems caused by the intended relation to ordinary language.

5.3.1. Nominal definitions

It is in arithmetic where the problems concerning definitions become important, because of the requirement that the defined things must always be classes (including the case of the definition of a single term, which was interpreted as the unique member of a class). Thus, in cardinals and ordinals we have (through the principle of abstraction) classes of classes or classes of series (which are classes as well). In both cases there is an underlying property respectively limiting the component classes: similarity and likeness. Sometimes it seems that the defined thing is a property in itself, as happens with infinity or continuity, but finally we always have classes (infinite classes and continuous classes —series). What characterizes its respective properties are always certain relations (e.g. a particular form of order). The same happens in geometry, where the things defined are also classes, either through the recourse to construct 'spaces' by transforming the axioms into parts of a definition (i.e. by assigning properties stating relations to certain objects), or through the direct interpretation of certain usual entities (e.g. straight lines) as mere relations with these or those properties.

As a whole, then, nominal definitions offered by Russell in practice intend to be limited to the field of the construction of classes by means of properties and relations. However, the wide generality obtained through this procedure does not manage *in fact* to overcome Peano's methods, especially because it rests on a principle of abstraction which (as we have seen) does not go beyond mere synonymy by offering only some techniques of conceptual economy. That is why Russell himself, in describing his method as based on definition, does it without being able to dispense with the ambiguity between classes, properties and relations, which can be hardly regarded as an alternative to Peano (POM, §474):

> a definition is always either the definition of a class, or the definition of the single member of a unit class: this is a necessary result of the plain fact that a definition can only be effected by assigning a property of the object or objects to be defined, *i.e.* by stating a propositional function which they are to satisfy.

The greatest ambiguity lies precisely in that only 'external' relations allowed Russell, in POM, to dispense with properties, whereas in stating that only classes can be defined, it seems that the possibility of defining relations (intensionally taken) is to be eliminated. The reason doubtlessly was that Russell regarded relations as definable only by means of their formal properties (symmetry, transitivity, etc.). After reading Frege he became convinced of the need to interpret them extensionally, i.e. as classes of couples. The problem was already pointed out by Jourdain in a letter of 1907, to which Russell replied the following (in Grattan-Guinness 1977a, 95):

> the dictum that all definitions are definitions of classes is certainly false. I was *meaning* 'no definitions are definitions of individuals'. But relations are just as definable as classes. Relations, however, being (in our calculus) relations in extension, may be regarded as classes of couples.

However, since we only can give an account of classes (and now of relations) by means of propositional functions, Russell adds that 'all definitions are definitions of propositional functions'. The return to the subject-predicate pattern was thus complete (in fact this return was underlying the earlier interest in the intuitiveness implicit in the vision of analysis as separation into parts).

This ambiguity is generated in the attempt of making Moore and Peano compatible. It is a correct thesis to say that Moore offered Russell the method of definitions based on the whole-part scheme, whereas Peano showed him how to carry it out in practice through concrete techniques. However, it requires bringing to light that the guarantee of the validity of definitions was, in every case, completely different. For Moore the basic criterion was the compromise between the direct intuition of the genuine simples and the meaning according to the ordinary language. Peano contributed with a constructive view of definitions which seemed to be compatible with Moore's, but this vision contained the seed of arbitrariness due to the interdefinability which was already implicit in the distinction between real and 'possible' definitions, as well as in the problem of implicit definitions (by abstraction or by postulates).

In a former section I already considered Russell's criticisms against definitions by abstraction. In fact the pre-eminence of nominal definitions already appeared in the beginning of *m1900c* (through the reference to Burali-Forti). However, the problem with such definitions was that, although they seemed to succeed in replacing mere vague intuitions by precise concepts, they were described by Peano as mere useful conventions, i.e. as recourses destined only to handle great groups of signs in an abbreviated way. On the other hand, the rigour required for its construction must rest on the Platonic intuition which was a guarantee of the reference to a real object. The problem was, then, the existing tension between two trends: according to the first one, it was necessary to *suppress* intuition and to replace it by logical rigour; according to the second, intuition was *preserved* just to be able to construct rigorous definitions. This tension remains worsened if we remember that, although Platonic intuition started from the realism of the 'third realm', it had available, as a practical guidance, only ordinary language (through the referentialistic theory of meaning). This led it to a different tension: the one existing between definitions which could be 'discovered', and the sense previously stated for the involved terms. The emphasis in realism led to the danger of breaking with the guidance of ordinary language, but this latter could lead, by itself, only to lexicography.

For Russell: 'An object may be present to the mind, without our knowing any concept of which the said object is *the* instance; and the discovery of such a concept is not a mere improvement in notation' (POM, §63). However, with that he did not allude to the true problem, for objects may be present to the mind only through language, and, on the other hand, they can be logically 'translated' only through the choice between different alternative formulas (as Peano showed with interdefinability). The referentialistic theory, inherited from Bradley, offered, on the other hand, the possibility of maintaining a strict correspondence between these two fields, by means of the intended parallelism between the logical form itself and language (once it has been modified). Starting from that it is necessary to admit, against Peano, that there are *true* and *false* definitions, and the criterion for the first ones lies in their correspondence with the objects at which we arrive through intuition. That is why Russell adds the following (POM, §63):

> as soon as the definition is found, it becomes wholly unnecessary to the reasoning to remember the actual object defined, since only concepts are relevant to our deductions. In the moment of discovery, the definition is seen to be *true*, because the object to be defined was already in our thoughts; but as part of our reasoning it is not true, but merely symbolic, since what the reasoning requires is not that it should deal with *that* object, but merely that it should deal with the object denoted by the definition.

With this latter distinction Russell does not deny that the concept constructed is referred to the object from the third realm, but simply affirming that denotation is ambiguous because it involves several variables which are referred to a whole class by means of a propositional function (see 4.2.1 above).

A similar problem takes place through the existing ambiguity between terms, classes and relations (see 4.4 and 5.2.2 above). On the one hand, the essentialistic needs led Russell to transform all definitions into nominal ones, which supposed that certain sets of axioms were 'absorbed' as parts of definitions; on the other hand, Russell denied that these axioms, by themselves, denoted unambiguously a genuine mathematical object. In other words: the philosophical needs led Russell to regard external relations as *ultimate*, which made it suitable to redefine everything in terms of relations; but relational structures (e.g. Peano's axioms) were rejected because they lacked the vision of the 'mind's eye'. In short: Russell was forced to defend, *at the same time*, the essentialism of terms (by declaring relations as secondary), and the essentialism of relations (by declaring terms as secondary: given a relation, its field is already stated). Moreover, relations are formulable as terms (on pain of falling into the extensionality of 'properties', admitted already in POM), and even as classes:[1]

$$xRy \;.\; \supset \;.\; x \in \rho \;.\; y \in \breve{\rho}.$$

5.3.2. Mathematical and philosophical definitions

Already in previous chapters I have considered the distinction between mathematics and philosophy. In 1.5 I pointed out Russell's criticisms against mathematical constructions in FG, due to his Kantian requirements of intuitiveness. Likewise, in 2.2 I showed his distrust towards the analysis of number (as an intuitive concept), which led him to reject the Cantorian constructions as 'merely' mathematical (see 2.8). The requirement of an intuitive ground was, therefore, constant in the entire stage before the Congress of 1900.

Russell's contact with Peano's powerful techniques to construct definitions produced a problem with deep ontological roots, since Moore's philosophy forced him to ontologically validate *all* concepts (i.e. terms), whereas Peano's definitions were presented as *eliminating* the defined concepts. It is true that Russell tried to overcome this problem by means of the identification of the logical indefinables with the ontological 'primitives', but the interdefinability supposed a limitation to this way. Logicist definitions were a good argument in favour of Moore since they seemed to reach the authentic indefinables of mathematics, but, on the other hand, they presented certain fundamental *concepts* of mathematics (e.g. the concept of number) near to an ontological disappearance. Russell's first reaction to this problem was prior to the acceptance of

---

[1] An instance of this situation is the possibility of regarding identity as either a relation, a property, a term, or a class (POM, §64).

logicism, and it already exhibited a certain distrust towards philosophical definitions, which, as Moorean analyses, were destined to bring to light the constituents of concepts, but without eliminating them. The enthusiasm for Peano's *Formulaire* leads Russell almost to reject such definitions (*m1900b*, 12):

> It should be observed that, in Mathematics, a term is considered to be defined when it is the only term having an assigned relation to one or more known terms. This is not the sense in which the word *definition* is usually used in philosophy; but it seems doubtful whether the philosophical use is capable of any precise meaning, and if it can be made precise, it would seem that, in the resulting sense, *all* ideas are indefinable.

This same idea is taken up in POM (§108) and, although Russell recognizes the unavoidable arbitrariness in choosing primitive notions, he emphasizes the *uniqueness* of the terms mathematically defined through the relation to one of those notions. On the other hand, Russell adds that philosophically the operation of defining is used only with regard to analysis as a division into parts, which is now rejected as inappropriate and useless. Perhaps Russell was impressed by the earlier analyses (in part I of POM) where he had verified again and again that if, following Moore, we regard the proposition as a complex concept equivalent to the *sum* of its constituents, the role of the logical form (i.e. order, structure, etc.) becomes lost. As we saw in 4.1.1 Moore's logic depended on the whole-part distinction (inclusion at bottom), which must be abandoned due to Peano's discovery of the membership relation, which allowed much more complex definitions and seemed to create mathematical objects as being unable to be embraced through pure extensionality (e.g. infinite classes). The rejection of philosophical definitions seems, then, to be definitive: 'I shall therefore, in future, ignore the philosophical sense, and speak only of mathematical definability' (*m1900b*, 12). However, Russell forgets in this moment his extreme need to open the way for a philosophical and intuitive foundation of all mathematical constructions, since these latter would not be able (in Russell's own view) to go beyond the framework of mere conventionality. Of course the problem became worsened when the reductive possibilities of logicism appear, and only the direct intuition of the genuine indefinables could avoid the bankruptcy involved in absolute arbitrariness.

In fact some forms of this incoherence have already appeared in the preceding chapter: the antithesis between ordinal and 'logical' properties (4.3); the defence of the construction of real numbers by 'philosophical' reasons (4.3); or the tension between formalism and 'intuitionism' in geometry (4.4). At bottom, what is underlying is the deep incoherence between the great technical possibilities of finding very different constructions, all of them equally valid from the operative viewpoint, and the need for presenting the formally admitted constructions as exhibiting the 'true meaning' of the defined notions.

Russell insists that mathematical definitions do not involve, like the philosophical ones, an analysis of the idea into its constituent elements (POM, §31):

> This notion, in any case, is only applicable to concepts, whereas in mathematics it is possible to define terms which are not concepts. Thus also many notions are defined by symbolic logic which are not capable of philosophical definition, since they are simple and unanalysable.

The difficulty with such a view is that (as happens in defining number one in terms of the unit class) the alluded mathematical definitions only provide a framework of relations, whose ground would remain obsolete if *philosophical* intuition does not designate the term actually pointed out by these relations: 'the mathematical definition does not point out the term in question... only what may be called philosophical insight reveals which it is among all the terms there are' (POM, §19).

Logicism, however, only could play the role of a strategy that should be formalist and intuitionistic at the same time, if the notion of *truth* as an indefinable is accepted (see 5.1.2), which already supposes a recourse to the intuition of something whose mathematical existence, although justified only through immediate inspection, obtains a definitive demonstration by means of the establishment of certain relations (of a structure). Therefore, the duality between mathematical and philosophical definitions could not become a true separation on pain of damaging the entire system. We have the best sign of this in the fact that even PM was built on this ambiguity, in spite of the fact that up until then Russell had already clarified some of the problems which had confused his usual presentation of logicism. The concise presentation of PM even made the incoherence clearer between the mathematical element of definitions as mere abbreviations and the philosophical element presenting them as genuine analyses of the considered ideas. This is nothing but (again) the attempt of taking advantage of the logical techniques without getting stuck with their 'tautological' disadvantages (which Russell tried to avoid from Couturat's suggestion to the contact with Wittgenstein; see my *1987a*, 351 ff).

Thus, we read in PM that definitions are abbreviations, typographical (and really superfluous) conveniences according to which it is stated that an introduced symbol would be equivalent to a group of other already known symbols (PM, I, 11-12):

> a definition is strictly speaking, no part of the subject in which occurs. For a definition is concerned wholly with the symbols, not with what they symbolize. Moreover it is not true or false, being the expression of a volition, not a proposition.

This view, which would transform definitions into nominal in the traditional sense of the term, is 'complemented' with another according to which definitions *may* be regarded as true or false since, in spite of being theoretically superfluous, 'they often convey more important information than is contained in the proposition in which they are used' (*ibidem*). However this is already referred to as something very different and closer to Aristotle's concept of 'real' definition (although I think it must not be understood exactly in the way followed by Weitz *1944a* and Robinson *1950a*, 195).

What happens is, rather, that Russell *does not admit the distinction*. As we have seen he intended that logic, through its indefinables, can serve for technical (mathematical) definitions and, at the same time, for intuitive (philosophical) definitions. That is why he justifies his second requirement in referring to the analysis of an idea into its constituents. Thus, when the defined idea is already familiar, 'the definition contains an analysis of a common idea, and may therefore express a notable advance' (PM, I, 12). With this, two completely different considerations are mixed: on the one hand, the assertion that mathematical definitions can be used to *analize concepts*; on the other, the underlying belief that this process coincides with the *precision* to be obtained from the ordinary language.

In Moore's philosophy both processes were not separated, and this non-distinction survives here as a fossil. In the regarded cases it is openly recognized that definitions give precision to a somewhat vague idea; with that, it certainly seems that the former attempt to relate definitions only to symbols is contradicted. However we must not forget that for Russell (since Whitehead renounced to the 'philosophical' part of PM; see Russell *1948a*), like for Moore, the separation between (ordinary or logical) language and the structure itself of our ontology is quite obscure, since only from language we can reach concepts, no matter how Platonic our vision of them can be. It is true that Russell at times speaks as if through a pre-linguistic intuition we can directly reach objects, but his constant care in preserving the connection with language supposes a certain softening of the apparent incoherence.

5.3.3. Analysis and ordinary language

The relation between analysis and ordinary language was present in Russell from the first contact with Bradley, since the old master was convinced of the need for reducing ordinary language to its true logical forms (see 1.4 above). Likewise, Russell made use from the beginning of his belief that certain concepts and axioms are *presupposed* in other more complex ones. In FG there was an attempt to find out such presuppositions in the field of geometry, although it contains as well a second deductive step (see 1.5 above). The manuscripts previous to the Congress of 1900 continue this line, though now by means of the notions of implication and logical priority, which are added to that of presupposition. The relation to language takes place from the need to discover the fundamental concepts through the different types of judgments, which can be isolated from ordinary (or mathematical) language. The involved implication is one based on inclusion (the whole-part) and is equivalent to the converse of logical priority (although it is simpler than this latter, which cannot be reciprocal) (see 2.4 above). This leads Russell to accept implication between concepts, and a view of analysis according to which it must look for the simplest concepts, which lack logical presuppositions (i.e. they are implied but do not imply anything). Therefore definitions, in being based on the intuition of the simple constituents, can be true or false, which made it easier to identify the logical and the ontological analyses, and both with the epistemological one.

However, already before the conversion to Peano's methods, Russell began to doubt about the validity of this 'logical order' based on a notion of implication that could be 'reciprocal' (see 4.3.4 above on *1901b*) and, therefore, gave rise to an unclear logical priority. The mastery of Peano's logic led him immediately to the rejection of the whole-part relation as a ground for logical calculus (see 4.1.1 above) and, with that, to the rejection of implication between concepts, remaining membership as the basic logical relation and the implication restricted to an indefinable relation between propositions (although already very close to the analysis in terms of truth values and, therefore, independent of the meanings of the involved concepts). This did not make the analysis invalid, but made it much more complicated, since it was possible to understand it (see 4.2.1 above) both as an 'empirical' analysis of logic until the discovery of a *sufficient* set of constants (Peano), and as a 'philosophical' analysis destined to show to the mind the undefinability of certain *necessary* fundamental notions (Moore).[1] As we shall see throughout this section, the attempts of making these two types of analyses compatible are the cause of the ultimate incoherence of presenting definitions as mere abbreviations and, at the same time, as analyses of the involved ideas. This, again, presents analysis as a genuine division into constituents and, at the same time, as a selection of the essential properties, i.e. as the process of giving precision to a somewhat vague notion. The first version will be more conceptual, whereas the second will become a practical process to improve our concepts.

However, the conception of analysis in POM avoided such problems and intended to progress in all directions at once. First of all, analysis is posed as a philosophical problem whose objective is to reach the simple concepts and axioms of mathematics, starting from the complex ideas involved (POM, §2). Of course such simple elements will belong to logic, so that its notions can be as well analyzed until obtaining the true indefinables, since analysis means definition: 'The method of discovering the logical constants is the analysis of symbolic logic' (§10). The 'proof' that the logicist analysis is the deepest one lies, according to Russell, in the reduction itself of arithmetic to logic. Peano's axioms, on the other hand, in admitting new indefinables for arithmetic, suppose 'a less degree of analysis' (§120), with which Russell seems to be defending a merely quantitative criterion.

Obviously the link between this method of analysis and ordinary language is the notion of 'true' meaning (POM, §§201-3). As there are true and false definitions (see 5.3.2 above), the only acceptable analyses will be those belonging to the first group; and as definitions also play the role of *stating* meanings, those being true will give the true meaning. However, the relation that this process should have with the already stated meaning of the term is not clear. Russell even says that the logical constituents of an idea constitute its meaning (§194), with which he seems to mean that the definition, if true, will identify the true meaning with the one already stated by this definition; that is to say, as if ordinary language had become different from the ideal logico-

---

[1] In fact, the ambiguity (of Peanian roots) between the necessary and the sufficient is present even in PM, where primitive notions are presented as *necessary* in the sense that if one of them is eliminated the whole system collapses (in the way it is built), as well as *sufficient* since there is no need to introduce any more notion to make all the stated deductions.

ontological language, and the mission of the philosopher was to make this difference lesser. With that Russell not only opposes Peano, who had a great respect for ordinary meaning (as the source of the properties of mathematical objects; see 3.1.3 above), but also Moore, for whom ordinary language should be always regarded as a philosophical guidance.

Russell was thoroughly aware of the danger of abandoning this firm ground, and he realized this mainly in those cases when the goal was 'to construct a definition' (POM, §276), rather than to pick up a more or less stated meaning, as it happens in the case of Cantor's definition of continuity, which is very far from the usual sense of the term. Russell protests for the supposed validity of this original sense, which he describes as mere Hegelian obscurity (§271). However he realizes that, in this case, the analysis seems to go beyond the guidance offered by language, in spite of the fact that Cantor's definition seems to him a 'triumph of analysis' (282), and he presents it as an authentic essentialistic discovery: 'This notion was presupposed in existing mathematics, though it was not known exactly what it was that was presupposed. And Cantor, by his almost unexampled lucidity, has successfully analyzed the extremely complex nature of spatial series' (§335). However he realizes, at the same time, that independently of the operative success of this construction, what Cantor achieved was an ordinal definition which was also completely general, free of contradictions and merely sufficient for mathematical purposes (§335),[1] with which, in lacking the *necessary* character, it seems doubtful that the construction can be regarded as genuine *analysis*.

I think the problem of elimination is underlying here, in the same way that it concerned the definition of number. If an object becomes reduced to its constituents when it is constructed, then it is eliminated and it seems that to preserve it as an important stage in the chain of definitions has no sense. What avoids this elimination is *precisely* the fact that this object has already an epistemological and linguistic status which makes the breaking with language impossible. Russell doubts that Cantor's continuity is very similar to the (vague) usual meaning of the term (POM, §282), but he cannot avoid the recognition that, if we accept the construction, it constitutes a firm ground that cannot be left aside because of merely 'verbal' problems.

However, Russell lacks a clear notion of the meaning of the term 'precision' and he seems to doubt between the stated and the constructed meaning. He realizes that precise definitions are necessary *even in philosophy* (this constitutes one of his more valuable legacies), for only they achieve that certain distinctions become more that mere arbitrary assumptions. That is why he thinks that Cantor's other important constructive definition, that of the infinite (coming really from Bolzano and Dedekind), showed that the supposed obvious distinction between finite and infinite was nothing but obscurity. According to that, it is undeniable that definitions *construct* their objects because they *appear* only through them: 'the distinction of the finite from the infinite is by no means easy, and *has only brought to light* by modern mathematics' (POM, §292).

---

[1] The lack of rigour, so usual at times in Russell, leads him even to recognize that, although the continuum is not analysable into indivisible elements, it can be divided by means of 'mathematical analysis'. With that he completely forgets his doctrine that analysis provides precisely the philosophical part of definitions; see his *1911d*, 286.

From the philosophical viewpoint there is no doubt that Russell concedes a great value to language (or to 'grammar'). For, starting from his belief (from this stage) that every word has some meaning, he was convinced that grammatical distinctions are a genuine guide for philosophical discoveries, and consequently that we can evaluate the correction of the analysis of a proposition by stating the meaning of its constituent words (POM, 46). In practice, however, it has sense when Russell limits himself to the attempt of classifying the ontology incorporated in the grammatical categories (substantive, adjective, verb, etc.) rather than when he makes use of them as an actual guidance for our constructions or as a criterion for the discovery of true 'logical forms'. In the Bradleian tradition, Russell regarded language as a misleading guide for these purposes (§72), and, as we saw above (especially in 4.2), what he did first was to analize language until finding the genuine logical forms, so that this analysis cannot be taken, at the same time, as a criterion for doing the same thing. That is why at bottom Russell always doubted between the respective pre-eminence of philosophical or mathematical definitions.

To reject substance (organic 'wholes'), which was the main ground for monism, Russell (and Moore) needed to reduce every form of unity to the whole formed out of its parts, which led him as well to erase the distinction between analysis as a mere conceptual process and genuine ontological division into parts (POM, §439), on pain of transforming the first one into something purely subjective. On the other hand, logic led him to recognize the existence of complex unities (propositions) not reducible to their constituents, on pain of losing their 'form'. Russell's solution was the ontological distinction between being and mere existence: 'all unities are propositions or propositional concepts... consequently nothing that exists is a unity. If, therefore, it is maintained that things are unities, we must reply that no things exist' (§439). What can be achieved in this way is only to reduce unities to the scheme of proposition, which was already characterized by the renunciation to the relational theory of judgment: as we saw above (in 4.2.1) the admission of propositional functions (here through propositional concepts[1]) is nothing but the surrendering to the distinction between term and concept (rejected by Moore).

With that the distinction between a proposition and any other complex concept is admitted, which rests, again, on the need for transforming properties into new concepts, but at the price of breaking, in practice, with this same scheme as a starting-point. Thus, even within the Platonism of the third realm, two incompatible theories of judgment are to be maintained. According to the first one, there is only one type of concepts, so that the analysis is always a division of complexes whose elements are located at the same ontological level; according to the second, only subjects are genuine concepts (terms), whereas adjectives cannot avoid becoming mere predicative appendices. However, although the whole problem is disguised under the language of propositional functions, the monistic danger cannot be avoided, i.e. the renunciation to maintain relations as not reducible to properties (the final adopted procedure by technical reasons). The defence of these two theories of judgment even became literal: 'There are two kinds of constituents in any complex: there are the terms and the relation joining them; and there can be (it

---

[1] In POM (§55) propositional concepts are introduced as the result of transforming a verb into a name (a logical subject). As we shall see this nominalization fully affects the same difficulty alluded here.

is possible) one term and the predicate qualifying it' (*1911a*, 56). As this distinction could not be made within Moore's philosophy, the influence of this author became increasingly small, although curiously Russell continued to maintain *external* relations even after PM, where relations were already not genuine entities.[1]

5.4. Relational logic and ontology

Russell insisted in POM that only Moore's philosophy could constitute an acceptable ontological ground, but he summed up this philosophy into two points: the non-existential theory of judgment, and pluralism (POM, preface), according to which the world must be regarded as an infinite number of independent entities whose (external) relations are irreducible to properties. With that he forgot the most important thing: to emphasize that this kind of pluralism is possible only by starting from the *impossibility of admitting predicates* being more than mere linguistic appearances without ontological implications. For only placing subject and predicate at the same level is it possible to regard relations as genuinely external, so avoiding the danger of transforming them into properties because of a simple reason: properties cease to exist.

The resulting ontology (POM, ch. 4) is much more complex than that of Moore, because it has to embrace the grammatical categories which serve as a guide to logic. However, although the general notion of 'term' (parallel to Moore's 'concept') is preserved as any thing than can be thought or can be a part of a proposition (as well as its character of immutable 'logical subject'), Russell immediately contradicts this presentation in dividing them into two types: things and concepts, from which only the first ones can be regarded as actual logical subjects of propositions. Thus, the recognition of logical subjects becomes a mere compliment towards Moore,[2] since they are reduced to the possibility that every term can be transformed into a logical subject of *certain* propositions (e.g. 'a logical subject is one'; §47).

This leads Russell to replace the notion of logical subject by that of 'proper name', coming from the Aristotelian logic (and rejected even by Bradley). As for concepts, they are identified in practice with properties, in being divided into predicates and relations. It is true that only the first ones are 'adjectives' (class-concepts) and therefore they can directly denote without the need for affirming anything, whereas relations (verbs, in general) automatically yield assertions (which is related, again, to the notion of truth) (§47). I think, however, that this analysis leaks through in

---

[1] In fact the separation from Moore began when the subject-predicate relation (still present in Russell *m1899a*) ceased to be seen as such a relation in next works.

[2] Russell admits in a note (POM, §48, 3) that his terms are different from Moore's concepts in several important points, but he does not draw the conclusions from this separation (in fact he did not turn to mention it).

many places. Firstly, concepts as adjectives are always predicates of some subject, so that the notion of truth is also unavoidable; and this has the consequence that the subject-predicate propositions can be regarded as similar to relational propositions, since both can be analyzed into subject and assertion (Russell admits this in §48). Besides, relations not being verbs (e.g. logical connectives) can also be presented as mere constituents in the denotation of a complex subject without the need for asserting anything. Therefore, a first approach leads us to the conclusion that it does not seem that the distinction holds, since two kinds of concepts as properties can be presented (and that without mentioning the fact that intransitive verbs directly give rise to properties[1]).

Russell seems to realize this is a very unsatisfactory situation when he tries to move away from the grammatical scheme, which leads him so dangerously to traditional philosophy. It is obvious that if a deep distinction between things and concepts is recognized, then it is necessary to admit a higher ontological status for the first ones, which would lead it to the old theory of substance and attributes. To avoid it, Russell presents the difference as merely grammatical (§49), arguing that if the framework of external relations is surpassed, then every adjective can be transformed into a substantive (a subject) with no change in the meaning, on pain that propositions referred to them are all false. Therefore, he adds, terms being concepts only differ from terms not being concepts because they appear in a different way in propositions. However when he tries to explain this different way of appearing, Russell declares it as 'indefinable' and proceeding from relations merely external (i.e. relations which do not affect the nature of the terms).

I think that the whole reasoning is wrong. It rests on a theory of judgment (that from Moore) precisely to defend the opposite one. According to the second one, he distinguishes between logical subjects and predicates on resting on the grammatical difference; according to the first, he explains this difference in terms of external relations. But speaking about external relations is sound only if all grammatical categories are overcome; however, Russell seems to speak as if external relations were merely 'apparent', whereas relations are true relations only it they are independent of terms and they acquire the status of new 'terms', which is precisely what makes them 'external'.

A sign that Russell continues to employ the relational theory is his criticism of Bradley, under the argument that propositions do not contain words, but 'entities indicated by words' (POM, §51). However, this can only be possible if we permit that all words can be logical subjects, with which relations merely grammatical are overcome: now they can be no longer 'external' since these latter relations only take place between the entities (in themselves) in propositions, but not between their equivalent words in ordinary language. On the other hand, Russell rests on the relational theory to build classes out of the corresponding predicates (class-concepts), which leads him, in the last analysis, to admit them together with the terms in themselves (things).

We have the same situation concerning the second kind of concept, verbs or relations (POM, §52). Also here Russell tries to dispense with grammatical differences which appeared through the

---

[1] Russell mentions such verbs (POM, §48), but only to add that in these cases the related term is 'indefinite'. I shall return below on this matter.

possibility of regarding the verb as a substantive, reducing them to 'merely external relations'.[1] Here a difficulty (already present in the former case) appears: in intransitive verbs the operation of transforming them into the corresponding substantives seems to eliminate the assertive character of the resulting complex (e.g. 'Caesar died' and 'the death of Caesar'). Russell had not yet available his later theory of descriptions, so that he could not carry out the corresponding analysis. On the other hand, he realizes that this assertive character does not depend on the grammatical form, as it is shown in the example '"*Caesar died*" is a proposition' (§52). However, this strengthened his belief that to avoid the transformation of certain entities into logical subjects is impossible, which is precisely what he needed to demonstrate the rigour of the distinction between things and concepts.

As for the theory of truth, it was left aside in POM (excepting the allusions we saw in 4.2.1) and it was not considered until *1904a*. However, when it appears in this work, it does so in a parallel way to the theme of the ontology of relations; in particular through the problem of whether relations are universal or can take place as particulars (which, again, is closely related to the theme of the subsistence of being). We might explain the global situation in the following way (although Russell seems to avoid such a direct explanation). For a logical theory of judgment (i.e. for a theory which does not identify it with mere belief), judgment is equivalent to proposition, i.e. to something being true or false;[2] therefore a wrong judgment points out a false proposition, which is its object. On the other hand, Russell's ontology conceded *being* even to non-existing things: false propositions must *subsist*. His logic had identified relations and verbs because they give rise to the unity of propositions; that is why as every proposition contains at least one verb, it contains also one relation. All that leads him to admit, as a criterion for the truth or falsehood of a proposition, the relation itself, which gives *being* to the proposition. However, if what is affirmed in a proposition is the involved relation, this will have to be a *particular* relation (since as universal relation it cannot be a doubtful entity) concerning a *concrete* instance. This particular relation cannot serve as a mark (Foster *1984a*, 140-1) to distinguish false propositions from the true ones because, as everything subsists, what is false has also being (*1904a*, 49):

> every constituent of a proposition, whether this proposition be true or false, must have being; consequently, if the particularised relation is a constituent of the proposition in which it is supposed to occur, then, since such a proposition is significant when it is false, the particularised relation has being even when the terms are not related by the relation in question.

---

[1] POM, §52. The position is, again, clearly anti-Bradleian: 'The question is: what logical difference is expressed by the difference of grammatical form? And it is plain that the difference must be one in external relations'. Thus, we come back to the mentioned vicious circle.

[2] Like in Moore, perception (the epistemological equivalence of judgment) always points out propositions, and only these latter (and not beliefs) exhibit truth or falsehood as properties.

Consequently, what is affirmed in the proposition is not the *being* of the supposed particular relation, so that the object of the judgment is not this relation. Only the proposition as a whole is the true object of the judgment, and in it only the *universal* relation is possible. Thus, by eliminating the only possible criterion to distinguish false propositions from the true ones, there remains no other way out that pure 'intuitionism': 'some propositions are true and some false, just as some roses are red and some white' (*1904a*, 75). With this (Moorean) theory the *dual* relation between the mind and the proposition as a whole is preserved.

On the other hand, the argument from POM against particular relations (§55) was less convincing in avoiding the theme of falsehood and considering a concrete relation: difference. However, it would bring to light the theme of identity and the principle of abstraction, which is interesting because of the connection with relational ontology. The argument is the following: although the difference between two terms is a particular relation, it cannot serve to reconstruct the proposition when we limit ourselves to enumerating (analyzing) its constituents, just like the universal relation would not serve. On the other hand, the fact that the differences are particular for each case (and that, consequently, they differed among themselves) would not avoid their having also something in common by virtue of which they could continue to be what they are (i.e. differences). At this point Russell applies the principle of abstraction: to have something in common is nothing but to be in a certain relation to a given term. Hence if no pair of terms can have the same relation, it follows that no pair of terms can have anything in common, so that 'different differences will not be in any definable sense *instances* of difference' (§55). The conclusion as for the theme of the difference is that the asserted relation between A and B in 'A differs from B' is nothing but the general relation of difference. The general conclusion is that relations have no particular instances: they are the same (even numerically) in every proposition where they are present.[1]

The problem with Moore is, again, unavoidable, since for him universals have only numerically diverse instances (see my *1990f*). However the interesting thing about this problem lies in that it arises from a common ground. For both of them any similarity can be always reduced to the possession of a relation to a third term (in fact, as we saw in 5.2 Moore already knew a version of the principle of abstraction). However, Russell tended to interpret this relation in terms of *predication*, pointing out that all terms having a certain relation to a third term form a class (POM, §96). Thus (like with the admission of 'things') he moved away from Moore's viewpoint according to which predication cannot be admitted, and that in spite of the fact that Russell, in some places of POM, still claimed that there is no essential difference between subject and predicate.[2]

It has been claimed (Griffin and Zak *1982a*, 61) that for Russell the transformation of a proposition into a propositional concept (e.g. 'Caesar died' and 'the death of Caesar') has the

---

[1] In fact all that has been demonstrated is that certain relations cannot be particular, but not that all of them are universal (see Griffin and Zak *1982a*, 57). To arrive at this latter position the support of the notion of truth and the subsistence of the false were needed (as we have seen).

[2] As may be seen in POM, §428, although here we find earlier ideas.

result of admitting the same *entity* and, therefore, the same proposition. If one applies this (we are told) to '*A* differs from *B*' one would obtain 'the difference of *A* and *B*', with which we would have the same proposition. Thus, by interpreting the relation of difference as a particular, the original proposition could be reconstructed. I think, however, that this interpretation would be objected to by Russell, and not only because in the case of the mentioned transformation the assertion is lacking (as the authors recognize), but mainly because Russell insists that a relational proposition refers to its terms (the terms of the relation) in a completely different way from that according to which it is referred to the relation itself. Thus there is a logical *actual* difference between the instance where the relation appears as such a relation, and the instance (the nominalization) where it is regarded as one more term (Russell *1904a*, 53-4).

Consequently, it is necessary to recognize that, again, Russell rests on the ambiguity between terms and relations. On the one hand he needs, due to his Moorean ontology, to regard everything as a term, including relations, which now have to be regarded as 'things'; on the other, he also needs to distinguish relations and things to build a relational ontology making pluralism possible. Finally, external relations, intensionally regarded, already carry the seed of the ambiguity since they can be confused with terms themselves. I am thinking, of course, of the doubt between the relation itself and the 'relating' relation, which leads one to Bradley's objection against the relation between relations and their terms.[1]

The evolution itself of this theme in terms of 'complexes', or the defence of external relations even after Russell's having admitted extensionality, are other signs that the ambiguity term-relation is a basic ingredient of Russell's ontology (at least until he explicitly admitted the replacement of inferences by logical constructions). I shall consider, to finish, these two implications in order to evaluate their contribution to Russell's definitive (implicit) rejection of Moore's philosophy.

In *m1905* Russell still maintains Moore's view that truth is a property of the objects of judgments, and that the proposition is a relation between several terms. Besides, he includes an attack against the theory of truth as correspondence with facts, with the argument (usual from Moore) that to state this theory one already makes use of truth. The conclusion is that the theory that truth is a property of judgments must be regarded as 'idealistic'. In *m1907?* Russell preserves the essentials of this view, is spite of the fact that he had already considered in detail the convenience of admitting a 'multiple' alternative theory in a technical manuscript of 1906 devoted to paradoxes (see my *1989a*). However, this manuscript seems to emphasize the fact itself of the belief, in stating the distinction between true facts and true beliefs (as a reaction against pragmatism, which made the distinction almost impossible). Finally, however, the canonical earlier position is admitted, although with the dangerous distinction between 'facts' and 'fictions' (instead of the simpler distinction between true and false propositions), perhaps being influenced by Meinong's use of fictions: 'Facts are true, and fictions false...; they are *objects* of beliefs and disbeliefs' (*m1907?*, 11). Of course, the problem was the admission of 'objective' fictions.

---

[1] POM, §§54 and 99 (see also 4.2.2). Griffin and Zak *1982a* also point this out, but they do not introduce it in the discussion, which seems to me not to be well linked to Russell's philosophy at this stage.

The first publication on this theme[1] (*1907a*) introduces for the first time what will be the seed of a progressive separation between ontology, logic and epistemology: the theory of judgment as a multiple relation. Now judgment is regarded as a belief, whereas it possesses objectivity only as a proposition. In this way the admission is avoided of the supposed 'objective' falsehood to which wrong beliefs would lead. The essential element of the change is the replacement of the belief as a 'single state of mind' by a set of 'several interrelated ideas' (*1907a*, 46). In this way the truth as correspondence can be admitted, since the correspondence between these 'ideas' and the 'complex' being its object would determine the truth or falsehood of the belief.

Russell realizes the difficulty of giving an account of this correspondence (in fact he admits he does not know how to define it; *1907a*, 46-7), but finally he decides to eliminate the supposed 'objective non-facts' and to emphasize the advantage of the new theory in allowing the distinction between perception and judgment (which was impossible in Moore's philosophy and even in Bradley's). Now one arrives, through perception, at the objective complex (the 'single state of mind'), and through judgment at a complex of 'presentations' (the *multiple relation*), whose constituents are identical with those of the objective complex, although they can be related in a different way.

With that Russell emphasizes the distinction between intuition and discursive knowledge, which would contribute to explain the 'infallibility' of perception (which is regarded, however, as somewhat doubtful; *1907a*, 49), although by explaining error through the subjectivity of judgment (*we* do state the multiple relation among the involved 'ideas'). The reason for the change must have been the existence of meaningful expressions without reference, and the possibility of eliminating certain false constituents (both through the theory of descriptions), as well as the logical problems involved in the need for dispensing with paradoxes (see my *1989a*). However the article arrives at its end by only regarding the new theory as a *possible* alternative with many difficulties.

The subsequent evolution of this theory (very well described in Griffin *1985a*) cannot be considered here in detail. It will be sufficient to say that the diverse presentations[2] achieved a greater clarity, mainly through dispensing with the first recourse to the 'ideas' and considering with much more care the objectivity of the involved complex. In PE it is clearly stated that judgment 'is not a dual relation of the mind to a single objective, but a multiple relation of the mind to the various other terms with which the judgment is concerned' (PE, 155), and in PM Russell even arrives at the rejection of propositions as 'incomplete symbols' which require the judgment itself (which is something subjective) to be able of having a meaning (PM, I, 44; see my *1989a*). This obviously supposed a breaking with Moore's philosophy, but it is interesting to

---

[1] There had been another earlier publication (*1906b*), but it was very brief and contained no detailed discussion of the involved problems; in fact it was only a first contact with Joachim's *The Nature of Truth*, ending with a global rejection (p. 533).

[2] Mainly PE (pp. 155-8), PM (I, 41-4) and PP (pp. 55-6). TK already offers elements which served for dispensing with the theory, in part due to Wittgenstein's criticisms (I considered all this problem in a fuller way in my *1990d*).

consider with some detail the way in which the involved subjectivity affects the relational ontology.

Russell makes efforts to avoid such a subjectivity by claiming that, although this strategy regards truth as a property of judgment (belief), and therefore depending on the existence of human minds, however, as the complex does not contain the judging mind, there is no subjectivity (PE, 158). However, the problem lies in the fact that the relation joining the constituents of this complex proceeds precisely from the judging mind, which, although not being a part of the complex, it *is* a constituent of the multiple relation of judging itself (PP, 56). In this way the *dual* relation, that before 1905 was regarded as the one existing between the mind and the proposition *as a whole*, now remains replaced by the relation of *believing*, which is multiple and takes place among several constituents in such a way that it can *impose* an ordering (a sense) not really existing. Consequently, *acquaintance* remained reserved to perception and, although it is 'infallible', it does not manage to capture the objective relations among the elements of a complex. The new subjectivity threatens, then, the already problematic tension between terms and relations we are considering (in the form of an ambiguity).

If it is admitted that in a complex one only accedes to the objective constituents, whereas relations depend, in some way, on ourselves, it seems we have again the old Bradleian theory of the tricky character of relations. The doubtful thing is, therefore, whether relations must or must not have the pre-eminence over the related terms. Winslade *1970a* has claimed that for Russell *relata* are ontologically previous to relations with the arguments that, if there are no relata, there will also be no relations. The thesis is doubtful and, although some passages are offered as evidence (which proceed mainly from a more 'linguistic' Russellian stage), I think that Winslade forgets Russell's great insistence that relations *create their own field* (see 4.2.2 and 4.4). The interesting thing in this article is, however, that it poses the problem of how the objectivity of relations is possible (they must be 'facts') if, as we saw above, the existence of particular relations is rejected, since only these relations would be *given* as such in the world.

This is also related to the theme of predicates, as Russell has insisted that these entities are similar to relations in that they cannot be particularized (*1944b*, 684). The truth is that in POM this particularization is claimed *precisely as the trait that distinguishes them from relations* (POM, §55). With that Russell was defending his recent discovery that the unity of the proposition depends on the verb (the relation). However, this same discovery provided no explanation of the fact that relations *as terms* can be a part of certain complexes (propositions) without that their *connection* with the rest of constituents is, again, explained (and that without mentioning the new problem having arisen through the return to the subject-predicate pattern —implicit in propositional functions— with regard to the breaking with the role of relations as elements 'synthesizing' two concepts located at the same ontological level).

I come now to the second announced theme: the consideration of relations as ultimate constituents of the world. It is easily understood that, in the same line of the manuscripts from 1898-1900, Russell defended external relations in POM with the argument that only through them can we overcome monism (or monadism). In fact in this work the link between externality and intensionality is complete, hence it is easy to regard relations as ultimate (POM, §216). But if one admits that relations are only ordered pairs (see 4.2.2 above), then Russell's insistence in maintaining the importance of external relations without the intensional support is hardly understood.

However the best attack against internal relations is that of *1907a*. I cannot study here its arguments, which are well-known, but I shall consider certain scattered allusions that can serve as illustrations of the tension between terms and relations. The 'axiom' of internal relations says 'every relation is grounded in the natures of the related terms' (*1907a*, 139). According to Russell, one of the consequences is ontological monism, which admits only a unique thing, since the supposed diversity could not be reducible to a difference of adjectives, i.e. to the natures of the involved terms. Thus, relations would be only adjectives of their terms. Russell concludes that 'when the axiom is rejected, it becomes meaningless to speak of the "nature" of the terms of a relation' (*1907a*, 146), but at this point he is implicitly accusing himself of having admitted the axiom, since he had claimed that certain entities (e.g. numbers) are 'constructed' and, however, they enjoy of a completely objective 'nature' in the Platonic sense of the term (e.g. POM, §427).

I shall consider, to finish, the evolution of the theme of particulars, which can illustrate the whole problem very well. The definitive breaking with the relational theory of judgment had taken place already in 1912. However, Russell presents his *1912b* as a coincidence with Moore *1900a* with the argument that there particulars are also admitted! The truth is that Moore introduced particulars only to defend the pre-eminence of numerical difference over conceptual difference, the possibility of particularization, and the rejection of the identity of indiscernibles (as I show in my *1990f*). However, all that started from the rejection of the subject-predicate pattern as a characteristic *relation* presupposing a *difference* between two types of concepts.

On the other hand, Russell *1912b* introduces the subject-predicate pattern as a full relation, presenting the existence of particulars as depending upon this same relation, whereas in the earlier stage he had denied that this pattern could be regarded as a relation.[1] As we saw in the beginning of this section, the division between things and concepts (in POM) implicitly contained this later development. What Russell is looking for now is to make this division explicit: if there is no relation of predication, we are told, there are no predicates and then particulars and universals are the same thing (*1912b*, 109). Curiously, the logical identification between particulars and logical subjects is so presupposed, although this would rather be the thesis to be demonstrated. However this view has the advantage of reducing the matter to its logical basis: 'if predication is an ultimate

---

[1] In POM Russell rejected subject-predicate pattern to be relational: 'In fact, subject-predicate propositions are distinguished by just this non-relational character' (§53; also in §428). Sprigge *1979a* is then mistaken when (in an effort to connect Russell's theory of relations to Bradley) he says that the argument of 'diversity' can also be applied to particulars and universals, adding that the first ones can be distinguished by their qualities and the second by themselves. Russell at times speaks in a way that produces this impression, especially when he surrenders to the subject-predicate pattern and states a dualistic ontology (things concepts) on logical grounds. However, in POM there are passages (like §428, p. 451, ch. 51) where Sprigge's thesis is literally rejected.

relation, the best definition of particulars is that they are entities which can only be subjects of predicates or terms of relations, i.e. that they are (in the logical sense) substances' (*1912b*, 123). It was the definitive surrendering to Peano against Moore.

The story, certainly, does not end here. In fact the evolution of Russell's ontology still suffered another great change towards simplification when he dispensed with particulars. But this depends on the complete explicitation of the method of logical constructions, and this, again, requires the detailed explanation of two previous steps: the influence of Wittgenstein and the assimilation of Whitehead's constructive techniques. Only by having explained the form in which Russell accepted the principle of Cournot-Couturat (see 1.3 above) and the way according to which he formulated a whole theory of ontological elimination through his theories of descriptions and types (see my *1989a*, *1989b* and *1990h*), one would be able to analyze Russell's main constructions from 1914 on, to show the deep unity of his philosophical method. However this task does not belong to this book (I tried to did it in my *1987a*, chs. 12-14, and my *1990i*), which has been devoted only to study the origins and the problems involved in Russell's mathematical philosophy as the main source of his subsequent philosophical method.

Bibliography

All unpublished manuscripts, together with the unpublished correspondence, are contained in *The Bertrand Russell Archives*, McMaster University, Hamilton, Ontario, Canada. I distinguish between actual manuscripts (ms) and typescripts (ts) (these latter generally made by the Russell Archives staff), and I quote them either by the folio number or (at times, especially in the case of major works) according to the *part* and the *chapter* of the work (e.g. IV/2), in order to avoid problems and preserve the references when these works become published. Besides, I offer the approximate number of folios of each manuscript. As for the published material, page numbers are generally referred to the last reimpression (or translation) listed here, although at times I mention a later reimpression only for the sake of completing the information (in these cases I do not include the pagination).

B.1. Works by Russell

B.1.1. Unpublished manuscripts

m1898   (AMR) *An analysis of mathematical reasonings*, 107 f, ts and ms.
m1898a   'Note on order', 29 f, ms (it includes material probably of 1899- 1900).
m1899   (FIAM) *The fundamental ideas and axioms of mathematics*, 59 f, ts and ms.
m1899a   'The classification of relations', 16 f, ts.
m1899b?   'On the principles of arithmetic', 23 f, ms.
m1900   (POM1) *Principles of mathematics*, 351 f, ms (20 f as ts). (In the first folio we can read: 'Draft of 1899-1900'.)
m1900a   'On identity', 1 f, ms.
m1900b   'Recent Italian work on the foundations of mathematics', 16 f, ts.
m1900c   'On the logic of relations', 59 f, ms (partially incorporated to the next one).
m1901?   'Sur la logique des relations', 31 f, ms (original unfinished version of *1901b*).
m1902?   'Necessity and possibility', 18 f, ts.
m1905   'The nature of truth', 24 f, ts.
m1907?   'The nature of truth', 11 f, ts.

B.1.2. Published or unpublished correspondence

- with G. Moore: 1896-1905, unpublished.
- with L. Couturat: 1897-1907, unpublished. Only the few originals in McMaster, from Couturat's side; I have not yet seen the rest, contained in the Biblioteque de la Ville, La Chaux-de-Fonds, Switzerland. For a general vision see Schmid *1983a*. The passages related to Leibniz have been published in O'Briant *1979a*.
- with G. Peano: 1901-1912. Published in Kennedy *1975a* (only letters from Peano's side; the rest has not been preserved).
- with G. Frege: 1902-1912. Published in Frege *1976a*, pp. 130-170. A good general review in Bell *1984a*.
- with P. Jourdain: 1902-1919. Published in Grattan-Guinness *1977a* (a fundamental work to understanding the development of Russell's philosophy).

B.1.3. Published works

1893a	'Paper on epistemology, II'. *1983a*: 124-30.
1895a	Review of G. Heymans' *Die Gesetze und Elemente des wissenshaftlichen Denkens*. Mind **4**: 245-9. Also in *1983a*.
1895b	'Observations on space and geometry'. *1983a*: 256-65.
1896a	'The logic of geometry'. *Mind* **5**: 1-23. Also in *1983a*.
1896b	Review of G. Lechalas' *Études sur l'espace et le temps*. *Mind* **5**: 128. Also in *1983a*: 287-8.
1896c	'The *à priori* in geometry'. *Proc. Arist. Soc.*, o.s. **3**: 97-112. Also in *1983a*.
1896d	Review of A. Hannequin's *Essai critique sur l'hypothèse des atoms dans la science contemporaine*. *Mind* **5**: 410-17.
1897a	(FG) *An essay on the foundations of geometry*, Cambridge: University Press.
1897b	Review of L. Couturat's *De l'infini mathématique*. *Mind* **6**: 112-9.
1897c	'On the relations of number and quantity'. *Mind* **6**: 326-41.
1898a	Review of A.E.H. Love's *Theoretical mechanics*. *Mind* **7**: 404-11.
1898b	'Les axiomes propres à Euclide, sont-ils empiriques?'. *Rev. Mét. Mor.* **6**: 759-76.
1899a	Review of A. Meinong's *Ueber die Bedeutung des Weberschen Gesetzes*. *Mind* **8**: 251-6.
1899b	'Sur les axiomes de la géométrie'. *Rev. Mét. Mor.* **7**: 684-707.
1900a	(PL) *A critical exposition of the philosophy of Leibniz*, Cambridge: University Press. Second edition with a new preface, London: Allen & Unwin, 1937.
1900b	'L'idée de l'ordre et de la position absolue dans l'espace et dans le temps'. *Congrés Int. de Phil.*, Paris, 1900. Paris: Colin, 1901, vol. III: 241-77.
1901a	'Sur la logique des relations'. *Rev. de Mathém.* **7**: 115-48. English trans. ('The logic of relations') by R.C. Marsch in 1956a: 3-38.
1901b	'On the notion of order'. *Mind* **10**: 30-51.
1901c	'Is position in time and space absolute or relative?'. *Mind* **10**: 293-317.

1901d	'Recent work on the principles of mathematics'. *International Monthly* **4**: 83-101. Reimp. ('Mathematics and the metaphysicians') in *1918a*: 75-95.
1901e	Review of W. Hastie's *Kant's cosmogony*. *Mind* **10**: 405-7.
1902a	'Geometry, Non-Euclidean'. *Encyc. Brit.*, 10th. edition, vol. 1: 664-74.
1902b	'On finite and infinite cardinal numbers'. Section III (pp. 378-83) of A.N. Whitehead, 'On cardinal numbers'. *Amer. Jrn. Maths.* **24**: 367-93.
1902c	'Théorie générale des séries bien-ordonnées'. *Rev. de Mathém.* **8**: 12-43.
1902d	'The study of mathematics'. Written in this year (see *1918a*, 5), but published in *New Quarterly* **1** (1907): 29-44. Reimp. in *1918a*: 61-74.
1902e	'What shall I read?'. Personal record of Russell's readings from 1891 to 1902 (not always reliable but useful), in *1983a*: 345-65.
1903a	(POM) *The principles of mathematics*, Cambridge (Univ. Press). Second edition with a new preface, London: Allen & Unwin, 1937.
1903b	'Recent work on the philosophy of Leibniz'. *Mind* **12**: 177-201.
1903c	Review of K. Geissler's *Die Grundsätze und das Wesen des Unendlichen in der Mathematik und Philosophie*. *Mind* **12**: 267-9.
1904a	'Meinong's theory of complexes and assumptions'. *Mind* **18**: 204-19; 336-54; 509-24. Also in *1973a*: 21-76.
1905a	'On denoting'. *Mind* **14**: 479-93. Also in 1956a: 41-56.
1906a	'The theory of implication'. *Amer. Jrn. Maths.* **28**: 159-202.
1906b	'The nature of truth'. *Mind* **15**: 528-33.
1907a	'On the nature of truth'. *Proc. Arist. Soc.* **7**: 28-49. Sections I and II ('The monistic theory of truth') in *1910b*: 131-146.
1908a	'Mathematical logic as based on the theory of types'. *Amer. Jrn. Maths.* **30**: 222-62. Also in *1956a*: 59-102.
1910a	(PM) (with A.N. Whitehead) *Principia mathematica*, vol I, Cambridge: University Press. Vols. II and III respectively appeared in 1912 and 1913. Second edition with new introduction in 1927.
1910b	(PE) *Philosophical essays*, London: Longmans Green. Second revised edition, London: Allen & Unwin, 1966.
1910c	'On the nature of truth and falsehood'. In *1910b*: 147-159.
1911a	'Le réalisme analytique'. *Bull. Soc. Franç. Phil.* **11**: 53-82. This is the transcription of a discussion among several people; the part by Russell in pp. 55-61.
1912a	(PP) *The problems of philosophy*, London: William & Norgate. I use the reimpression in Oxford: University Press, 1980.
1912b	'On the relation of universals and particulars'. *Proc. Arist. Soc.* **12**: 1-24. Also in *1956a*: 105-124.
1914a	(OKEW) *Our knowledge of the external world*, London: Allen & Unwin. Second edition with a new preface and some changes in 1929.
1918a	(ML) *Mysticism and logic*, London: Longmans Green.
1919a	(IMP) *Introduction to mathematical philosophy*, London: Allen & Unwin.

1924a	'Logical atomism'. *Contemporary British philosophy*, first series, London: Allen & Unwin: 356-83. Also in *1956a*: 323-43.
1937a	'Introduction to the second edition'. In *1903a*, 2nd. ed., London: Allen & Unwin: v-xiv.
1944a	'My mental development'. In Schilpp *1944a*: 3-20.
1944b	'Reply to criticisms'. In Schilpp *1944a*: 681-741 (see also *1971a*).
1948a	'Whitehead and *Principia Mathematica*'. *Mind* **57**: 137-8.
1951a	'Is mathematics purely linguistic?' (the year is doubtful). In 1973a: 295-306.
1953a	'Alfred North Whitehead'. *Riv. Crit. Stor. Filos.* **8**: 101-4 (transcription of a reading in the BBC). Reimpression in *1956a*.
1955a	'My debt to German learning'. In *1956a* as part of 'Other philosophical contacts'.
1956a	(PRM) *Portraits from memory and other essays*, London: Allen & Unwin.
1959a	(MPD) *My philosophical development*, London: Allen & Unwin. I use the reimpression in London: Unwin Books, 1975.
1967a	(AB1) *The autobiography of Bertrand Russell*, vol. 1 (1872-1914), London: Allen & Unwin.
1971a	'Addendum to my "Reply to criticisms"'. In Schilpp *1944a* (edition of 1971): xvii-xx.
1973a	(EA) *Essays in analysis*, D. Lackey (ed.), London: Allen & Unwin.
1983a	(CP1) *Cambridge essays, 1888-99*, vol. 1 of *The Collected Papers of Bertrand Russell*, London: Allen & Unwin.
1984a	(TK) *Theory of knowledge: the 1913 manuscript*, vol 7 of CP, London: Allen & Unwin.

B.2. Works by other authors

Aimonetto, I. *1969a* 'Il concetto di numero naturale in Frege, Dedekind e Peano'. *Filosofia* **20**: 580-606.
Barone, F. *1966a* 'Peirce e Schröder'. *Filosofia* **17**: 181-224.
Bell, D. *1984a* 'Russell's correspondence with Frege'. *Russell* **3**: 159-70.
Benacerraf, P. *1965a* 'What numbers could not be'. In Benacerraf/Putnam *1983a*: 272-94.
Benacerraf, P. and Putnam, H. *1964a* (eds.) *Philosophy of mathematics,* Englewood Cliffs, N.J.: Prentice Hall.
   1983a (eds.) Second edition of *1964a*, with many changes, Cambridge: University Press.
Blackwell, K. *1985a* 'Part I of *The principles of mathematics*'. *Russell* **4**: 271-88.
Bôcher, M. *1904a* 'The fundamental conceptions and methods of mathematics'. *Bull. Amer. Math. Soc.* **11**: 115-35.
Bonfantini, M.A. *1970a* 'Il primo Russell o il canto del cigno della geometria *kantiana*'. *Acme* **23**: 359-433.
Boole, G. *1847a The mathematical analysis of logic*, Oxford: Basil Blackwell, 1965.
Botazzini, V. *1985a* 'Dell'analisi matematica al calcolo geometrico: origine delle prime richerche di logica di Peano'. *Hist. Phil. Log.* **6**: 25-52.
Bourbaki, N. *1969a Eléments d'histoire des mathématiques*, Paris: Herman. Spanish trans., Madrid:

Alianza, 1976.

Bradley, F.H. *1883a The principles of logic*, 2 vols, Oxford: University Press (2nd. ed., 1922).

1893a *Appearance and reality*, Oxford: University Press (2nd. ed., 1897).

Bunn, R. *1980a* 'Developments in the foundations of mathematics, 1870-1910'. In Grattan- Guinness *1980a*: 220-55.

Burali-Forti, C. *1896a* 'Le classi finite'. *Atti dell'Accad. di Torino* **32**: 34-52.

1897a 'Una questione sui numeri transfiniti'. *Rendic. Circ. Math. Palermo* **11**: 154-64.

1899a 'Les proprétés formales des operations algébriques'. *Rec. Mathém.* **6**: 141, 77.

1899b 'Sur l'egalité et sur l'introduction des éléments dérivés dans la science'. *L'enseign. math.* **1**: 246-61.

1900a 'Sur les différentes méthodes logiques pour la definition du nombre réel'. *Congrès Int. de Phil.*, Paris, 1900. Paris: Colin, 1901, vol. III: 289-307.

Bynum, T.W. *1972a* 'On the life and work of G. Frege'. Introduction to Frege *1879a*: 1-54.

Cantor, G. *1883a Grundlagen einer allgemeinen Mannigfaltigkeitslehre.* ... Leipzig: Teubner. Partial French trans. ('Fondaments d'une théorie générale des ensembles'), *Acta Mathém.* **2**: 381-408. Partial English trans. by U. Parpart ('Foundations of the theory of manifolds'), *The Campaigner* **9** (1976): 69-96.

1895a 'Beiträge zur Begründung der transfiniten Mengenlehre' part I. *Mathematische Annalen* **46** (part II, 1897 **49**). English trans. by P. Jourdain (*Contributions to the founding of the theory of transfinite numbers*), Chicago: Open Court, 1915.

1937a Briefwechsel (Cantor, G. Dedekind, R.), J. Cavaillès and E. Noether (eds.), Paris: Hermann. French trans. (with some more letters) in Cavaillès *1962a*: 179-251.

Carnap, R. *1931a* 'Die logiszistische Grundlegung der Mathematik'. *Erkenntnis* **2**. English trans. ('The logicist foundations of mathematics') in Benacerraf/Putnam *1983a*: 41-52.

Cassina, U. *1933a* 'L'oeuvre philosophique de G. Peano'. *Rev. Mét. Mor.* **40**: 481-91.

1933b 'Su la logica matematica di G. Peano'. *Bull. Unione Mat. Ital.* **12**: 57-65.

1954a 'Sulle definizione per abstrazione'. *Atti I Congr. Stud. Metod.*, Torino: Ramella: 155-61.

1955a 'Storia ed analisi del 'Formulario completo' di G. Peano'. *Bull. Unione Mat. Ital.* **10**: 244-65; 544-74.

Cavaillès, J. *1938a* 'Remarques sur la formation de la théorie abstraite des ensembles'. In *1962a*: 25-176.

1962a *Philosophie mathématique*, Paris: Hermann.

Clark, R.W. *1975a The life of Bertrand Russell*, London: Penguin, 1978.

Couturat, L. 1896a *De l'infini mathématique*, Paris: Alcan.

1898a Review of Russell's *Foundations of geometry. Rev. Mét. Mor.* **6**: 354-8.

1898b 'Sur les rapports du nombre et de la grandeur'. *Rev. Mét. Mor.* **6**: 422-47.

1899a 'La logique mathématique de M. Peano'. *Rev. Mét. Mor.* **7**: 616-46.

1900a 'Sur une définition logique du nombre'. *Rev. Mét. Mor.* **8**: 23-36.

1900c 'Sur la définition du continu'. *Rev. Mét. Mor.* **8**: 157-68.

1900d 'L'algèbre universelle de M. Whitehead'. *Rev. Mét. Mor.* **8**: 323-62.

1900e 'Congrès Int. Phil. Paris 1900', sect. III. *Rev. Mét. Mor.* **8**: 555-65; 598-98.

1901a Review of G. Peano's *Formulaire... (1894-1901)*. *Bull. Sci. Math.* **25**/I: 141-59.

1904a 'Les principes des mathématiques'. *Rev. Mét. Mor.* (I) **12**: 19-50; (II-III) **12**: 211-40; (IV-V) **12**: 664-98; (VI) **12**: 811-44; (VI -cont.) **13** (1905): 224-56.

1904b 'La philosophie des mathématiques de Kant'. *Rev. Mét. Mor.* **12**: 321-83.

1904c 'Congrès Int. Phil. Geneve 1904', sect. II. *Rev. Mét. Mor.* **12**: 1037-77.

Couturat, L. et al. *1983a L'oeuvre de Louis Couturat (1868-1914)*, Paris: Presses de l'École normale supérieure.

Dauben, J.W. *1979a Georg Cantor*, Cambridge, Mass.: Harvard University Press.

1980a 'The development of Cantorian set theory'. In Grattan-Guinness *1980a*: 181-219.

de Lorenzo, J. *1974a La filosofía de la matemática de Poncaré*, Madrid: Tecnos.

Dedekind, R. *1872a Stetigkeit und irrationale Zahlen*, Brunswick: Vieweg. English trans. ('Continuity and irrational numbers') in *1901a*: 1-27.

1888a Was sind und was sollen die zahlen?, Brunswick: Vieweg. English trans. ('The nature and meaning of numbers') in *1901a*: 31-115.

1890a 'Letter to Keferstein'. In van Heijenoort *1967a*: 90-103.

1901a Essay on the theory of numbers, (trans. by W.W. Beman) Chicago: Open Court. Reimp. in New York: Dover, 1963.

Dieudonné, J. *1983a* 'Louis Couturat et les mathématiques de son époque'. In Couturat *et al. 1983a*: 97-112.

Dugac, P. *1976a Richard Dedekind et les fondements des mathématiques*, Paris: Vrin.

1983a 'Louis Couturat et Georg Cantor'. In Couturat *et al. 1983a*: 55-62.

Eames, E. R. *1969a Bertrand Russell's theory of knoweledge*, London: Allen & Unwin

Feinberg, B. *1967a* (ed.) *A detailed catalogue of the Archives of Bertrand Russell*, London Continuum 1 Ltd.

Foster, T. *1984a* 'Russell on particularized relations'. *Russell* **1**: 129-43.

Frege, G. *1879a Begriffsschrift,...* Halle: Nebert. English trans. by T.W. Bynum (*Conceptual notation ...*), Oxford: Clarendon, 1972.

1884a Die Grundlagen der Arithmetik. ... Breslau: Koebner. Spanish trans. by U.Moulines in Barcelona: Laia, 1972. Partial English trans. by M.S. Mahoney in Benacerraf/Putnam *1964a* and *1983a*: 130-159.

1893a Grundgesetze der Arithmetik, ... vol. I. Jena: Pohle. English partial trans. (Introd. and §§0-52) by M. Furth (*The basic laws of arithmetic*), Berkeley: University of California Press, 1964.

1976 Wissenschaftlicher Briefwechsel, Gabriel, G. *et al.* (eds.), Hamburg: Meiner. English partial trans. by H. Kaal (*Philosophical and mathematical correspondence*), Oxford: Blackwell, 1980.

Freudenthal, H. *1962a* 'The main trends in the foundations of geometry in the 19th. century'. In Nagel *et al. 1962a*: 613-21.

Garciadiego, A. *1985a* 'The emergence of some of the nonlogical paradoxes of the theory of sets, 1903-1908'. *Hist. Math.* **12**: 337-51.

1985b 'Una nueva interpretacion del origen del logicismo'. *Cuadernos de café y matemáticas* (Mexico) No 1: 59-66.

1985c 'L'influence de G. Cantor sur Bertrand Russell'. *Sem. Hist. Math. de Toulouse* **8** (1986): 1-14.

1986a 'On rewriting the history of the foundations of mathematics at the turn of the century'. *Hist. Math.* **13**: 39-41.

- *1987a* 'Russell's precise language'. Lecture to the *Canadian Soc. Hist. Phil. Math.*, McMaster University, Hamilton, Canada.
- *1987b* 'A biobibliographical note on Bertrand Russell's *The principles of mathematics*'. Unpublished article kindly provided by the author (together with the former one and his Ph.D. thesis: *B. Russell and the origin of the set theoretic paradoxes*, Toronto University, 1983).

Gergonne, M. *1818a* 'Essay sur la théorie des définitions'. *Ann. Math. Pur. Apl.* **9**: 1-35.

Geymonat, L. *1955a* 'I fondamenti dell'aritmetica secondo Peano e le obiezioni 'filosofiche' di Bertrand Russell'. In Terracini *1955a*: 51-63.

Gillies, D.A. *1982a Frege, Dedekind and Peano on the foundations of arithmetic*, Assen: Van Gorcum.

Gödel, K. *1944a* 'Russell's mathematical logic'. In Schilpp *1944a*: 125-53.

González, L. *1979a* 'Lógica y filosofía en Whitehead'. *Teorema* **9**: 299-322.

Grattan-Guinness, I. *1975a* 'Wiener on the logics of Russell and Schröder. An account of his doctoral thesis, and of his discussion of it with Russell'. *Ann. Sci.* **32**: 103-132
- *1976a* 'On the mathematical and philosophical background to Russell's *The principles of mathematics*'. In Thomas/Blackwell *1976a*: 157-173.
- *1977a Dear Russell - Dear Jourdain. A commentary on Russell's logic, based on his correspondence with Philip Jourdain*, London: Duckworth.
- *1978a* 'How Russell discovered his paradox'. *Hist. Math.* **5**: 127-37.
- *1980a* (ed.) *From the calculus to set theory, 1630-1910*, London: Duckworth. (There is also Spanish edition, Madrid: Alianza.)
- *1980b* 'The emergence of mathematical analysis and its foundational progress, 1780-1880'. *1980a*: 94-148.
- *1980c* 'Georg Cantor's influence on Bertrand Russell'. *Hist. Phil. Log.* **1**: 61-93.
- *1985a* 'Bertrand Russell's logical manuscripts: an apprehensive brief'. *Hist. Phil. Log.* **6**: 53-74.
- *1986a* 'Russell's logicism versus Oxbridge logics 1890-1925: a contribution to the real history'. *Russell* **5**: 101-31.
- *1986b* 'From Weierstrass to Russell: a Peano medley'. *Celebrazioni in memoria di G. Peano*. Università di Torino.
- *1988a* 'Living together and living apart. On the interactions between mathematics and logics from the French Revolution to the first World War'. *S. Afr. J. Philos.* **7/2**: 73-82.
- *1990a* 'Peirce between logic and mathematics'. Improved version of a paper read in the Symposium 'Peirce's contribution of logic', Cambridge, Massachusetts, September 1989. To appear in the proceedings.

Griffin, N. *1980a* 'Russell on the nature of logic (1903-1913)'. *Synthese* **45**: 117-188.
- *1985a* 'Russell's multiple relation theory of judgment'. *Phil. Stud.* **47**: 213-47.

Griffin, N. and Zak, G. *1982a* 'Russell on specific and universal relations'. *Hist. Phil. Log.* **3**: 55-67.

Hannequinn, A. *1895a Essai critique sur l'hypothèse des atoms*, Paris: Masson.

Harrell. M. *1988a* 'Extension to geometry of *Principia Mathematica* and related systems II'. *Russell* **8**: 140-160.

Hesse, M. *1952a* 'Boole's philosophy of logic'. *Ann. Sci.* **8**: 61-81.

Hilbert, D. *1899a Grundlagen der Geometrie*, Leipzig: Teubner. Spanish trans. from the 7th edition (1930) by F. Cebrián, Madrid: CSIC, 1953.

Imbert, C. *1969a* 'Estudio de los *Fundamentos de la aritmética* de Frege'. Spanish trans. of a French study (Paris: Seuil), in the Spanish trans. of Frege *1884a*: 130-238.

Ishiguro, H. *1972a Leibniz's philosophy of logic and language*, London: Duckworth.

Jager, R. *1972a The development of Bertrand Russell's philosophy*, London: Allen & Unwin.

Jørgensen, J. *1931a A treatise of formal logic*, 3 vols. New York: Russell, 1962.

Jourdain, P.E.B. *1906a* 'The development of the theory of transfinite numbers'. *Archiv der Mathematik und Physik* (I) **10** (1906): 254-81; (II) **14** (1908-9): 287-311; (III) **16** (1910): 21-43; (IV) **22** (1913-4): 1-21.

    1910a 'Transfinite numbers and the principles of mathematics'. *Monist* **20**: 93-118.

    1910b 'The development of theories of mathematical logic and the principles of mathematics'. *Quarter. Jrn. pure appl. Math.* (I) **41** (1910): 324-52; (II) **43** (1912): 219-314; (III) **44** (1913): 113-28.

    1912a 'Mr. Russell's first work on the principles of mathematics'. *Monist* **22**: 149-58.

    1915b 'Introduction'. In Cantor *1895a*: 1-82.

Kant, I. *1781a Kritik der reinen Vernunft*, Riga (2nd ed., 1787). Spanish trans. by P. Ribas, Madrid: Alfaguara, 1978.

Kennedy, H.C. *1963a* 'The mathematical philosophy of Peano'. *Phil. Sci.* **30**: 262-6.

    1973a 'The origins of modern axiomatics: Pasch to Peano'. *Am. Math. Month.* **79**: 133-6.

    1973b 'What Russell learned from Peano'. *Notre Dame Jrn. form. Log.* **14**: 367-72.

    1974a 'Peano's concept of number'. *Hist. Math.* **1**: 387-408.

    1975a 'Nine letters from G. Peano to B. Russell'. *Jrn. Hist. Phil.* **13**: 205-20.

    1980a Peano. Life and works of Giuseppe Peano, Dordrecht: Reidel.

Kilmister, C.W. *1984a Russell*, London: Harvester.

Klemke, E.D. *1970a* (ed.) *Essays on Bertrand Russell*, Urbana, Ill.: University of Illinois Press.

Kline, M. *1972a Mathematical thought from ancient to modern times*, Oxford: University Press.

Kummer, E.E. *1847a* 'On ideal numbers'. In Smith *1929a*: 119-26.

Largeault, J. *1970a Logique et philosophie chez Frege*, Paris: Nauwelaerts.

Lowe, V. *1941a* 'The development of Whitehead's philosophy'. In Schilpp *1941a*: 15-124.

    1985a Alfred North Whitehead. The man and his work. Vol. I: 1861-1910, Baltimore, Mar.: J. Hopkins University Press.

McColl, H, *1905a* 'Symbolic reasoning'. *Mind*, n.s. **14**: 74-81. Reimp. in Russell *1973a*: 308-16.

    1905b 'Three notes from *Mind* '. *Mind*, n.s. **14**: 295-6; 401-2; 578-9. Reimp. in Russell *1973a*, 317-22.

Mangione, C. *1972a* 'La lógica en el siglo XX' (Spanish trans.). In L. Geymonat (ed.), *Historia del pensamiento filosófico y científico*, Barcelona: Ariel, 1985, vol. II: 202-421.

Moore, G.E. *1899a* 'The nature of judgment'. *Mind* **8**: 176-93.

    1899b Review to Russell's *Essay on the foundations of geometry*. *Mind* **8**: 397-405.

    1900a 'Identity'. *Proc. Arist. Soc.* **1**: 103-27.

    1900b 'Necessity'. *Mind* **9**: 288-304.

    1903a Principia Ethica, Cambridge: University Press.

    1903b 'Experience and empiricism'. *Proc. Arist. Soc.* **3**: 80-95.

1903c   'The refutation of idealism'. *Mind* **12**: 433-53.
1942a   'An autobiography'.In Schilpp *1942a*: 1-39.
1942b   'A reply to my critics'. In Schilpp *1942a*: 533-676.
1952a   'Addendum to my reply'. In Schilpp *1942a* (2nd edition): 677-88.

Moore, G.H. and Garciadiego, A. *1981a* 'Burali-Forti's paradox: a reappraisal of its origins'. *Hist. Math.* **8**: 319-50.

Mosterín, J. *1978a* 'Sobre el concepto de modelo'. *Teorema* **8**: 131-42.

1980a   'La polémica entre Frege y Hilbert acerca del método axiomático'. *Teorema* **10**: 287-306.

Nagel, E. *1939a* 'The formation of modern conceptions of formal logic in the development of geometry'. *Osiris* **7**: 142-224.

Nagel, E., Suppes, P. and Tarski, A. *1962a* (eds.) *Logic, methodology and philosophy of science*, Stanford, Cal.: University Press.

Nakhnikian, G. *1974a* (ed.) *Bertrand Russell's philosophy*, London: Duckworth.

O'Briant, W.H. *1979a* 'Russell on Leibniz'. *Studia Leibnitiana* **11/2**: 159-222.

Padoa, A. *1899a* 'Note di Logica matematica'. *Rev. Mathém.* **6**: 105-21.

1900a   'Essai d'une théorie algébrique des nombres entiers, précédé d'une introduction logique a une théorie déductive quelconque'. *Congrés Int. de Phil.*, Paris, 1900. Paris: Colin, vol. III: 309-65.

1900b   'Un nouveau système irréductible de postulats pour l'algèbre'. *Compte rendu du Deuxième Congrès Int. des mathématiciens*, Paris, 1900. Paris: Gauthier-Villars, 1902, vol. II: 249-56.

Passmore, J. *1969a* 'Russell and Bradley'. In Brown and Rollins (eds.), *Contemporary philosophy in Australia*, London: Allen & Unwin, 1969: 21-30.

Peano, G. *Opere Scelte* (OS), 3 vols. Ugo Cassina (ed.), Roma: Cremonese: 1957-9.

1888a   'Operazioni della logica deduttiva'. OS2: 3-19 (partial).
1889a   *Arithmetices principia, nova methodo exposita*. Reimp. and Spanish trans. by J. Velarde, Oviedo: Pentalfa, 1979.
1889b   'I principii di geometria logicamente esposti'. OS2: 56-91.
1890b   'Sur une courbe qui remplit toute une aire plane'. OS1: 110-4.
1890c   'Démonstration de l'intégrabilité des équations différentielles ordinaires'. OS1: 119-70.
1891a   'Gli elementi di calcolo geometrico'. OS3: 41-71.
1891b   'Principii di logica matematica'. OS2: 92-101.
1891c   'Formole di logica matematica'. OS2: 102-13.
1891d   'Sul concetto di numero'. OS3: 80-109.
1891e   Review of Schröder's *Vorlesungen über die Algebra der Logik*. OS2: 114-21.
1894a   'Sui fondamenti della geometria'. OS3: 115-57.
1894b   'Notations de logique mathématique'. OS2: 123-75.
1894c   'Sur la définition de la limite d'une fonction. Exercici de logique mathématique'. OS1: 228-57.
1894e   Two letters to F. Klein. In Dugac *1976a*: 267-8.
1895a   'Logique mathématique'. OS2: 177-88 (partial).
1895c   Review of Frege's *Grundgesetze der Arithmetik*. OS2: 189-95.
1896a   'Introduction' to vol. II of the *Formulaire*. OS2: 196-200.

1896b   'Saggio di calcolo geometrico'. OS3: 167-86.
1897a   'Studii di logica matematica'. OS2: 201-17.
1897b   'Logique mathématique' (from vol. II of the *Formulaire*). OS2: 218-81.
1898a   'Sulle formule di logica'. OS2: 282-6.
1898b   'Risposta ad una lettera di G. Frege'. OS2: 295-6.
1898c   'Analisi della teoria dei vettori'. OS3: 187-207.
1898d   'I fondamenti dell'aritmetica nel Formulario del 1898'. OS3: 215-31.
1898e   'Sul paragrafo 2 del Formulario t. II'. OS3: 232-48.
1898f   Review of Schröder's 'Ueber Pasigraphie…'. OS2: 297-303.
1899a   'Sui numeri irrazionali'. OS3: 249-67.
1900a   'Formules de logique mathématique'. OS2: 304-61.
1900b   'Les définitions mathématiques'. OS2: 362-8.
1901a   *Formulaire de mathématiques*, Paris: Carré et Naud.
1901b   Review of Stolz's and Gmeiner's *Theoretische Arithmetik*. OS3: 360-3.
1906a   'Super theorema de Cantor-Bernstein ed Additione'. OS1: 337-58.
1908a   'La curva di Peano nel *Formulario mathematico*'. OS1: 115-6.
1911a   'Sulla defizione di funzione'. OS1: 363-5.
1911b   'Le definizioni in matematica'. *Arx. Inst. Ciències* (Barcelona) **1**: 49-70.
1912a   'Sulla definizione di probabilità'. OS1: 366-8.
1912b   'Delle proposizioni esistenziali'. OS2: 384-8.
1913a   'Sulla defizione di limite'. OS1: 389-409.
1913b   Review of Russell's and Whitehead's *Principia mathematica*. OS2: 389-401.
1915b   'Le definizioni per astrazione'. OS2: 402-16.
1921a   'Le definizioni in matematica'. OS2: 423-35.

Pears, D. 1972a (ed.) *Bertrand Russell. A collection of critical essays*, New York: Doubleday.

Peirce, C.S. *Collected Papers of C.S. Peirce* (CPP), vols. 3 and 4. Cambridge, Mass.: Harvard University Press, 1933. Reimp.: Belknap, 1967.
1867a   'Upon the logic of mathematics'. CPP3: 3-26.
1881a   'On the logic of number'. CPP3: 158-70.
1885a   'On the algebra of logic'. CPP3: 210-49.
1896a   'The regenerated logic'. CPP3: 266-87.
1897a   'The logic of relatives'. *Monist* **7**: 161-217.
1897b   'A theory about quantity'. CPP4: 132-44 (before unpublished).
1898a   'The logic of mathematics in relation to education'. CPP3: 346-59.
1902a   'The simplest mathematics'. CPP4: 189-262 (before unpublished).

Pieri, M. *1898a* 'I principii della geometria di posizione'. *Memorie Real. Accad. Sc. di Torino* **48**: 1-62.
1899a   'Della geometria elementare come sistema ipotetico deduttivo'. *Memorie Real. Accad. Sc. di Torino* **49**: 173-222.
1900a   'Sur la géométrie envisagée comme un système purement logique'. *Congrès Int. de Phil.*, Paris, 1900. Paris: Colin, 1901, vol. III: 367-404.

1906a 'Sur la compatibilité des axiomes de l'arithmétique'. *Rev. Mét. Mor.* **14**: 196-207.

Poincaré, H. *1899a* 'Des fondements de la géométrie. A propos d'un livre de M. Russell'. *Rev. Mét. Mor.* **7**: 251-79.

1900a 'Sur les principes de la géométrie'. *Rev. Mét. Mor.* **8**: 73-86.

Queysanne, M. *1964a Algèbre,* Paris: Colin.

Quine, W.O. *1941a* 'Whitehead and the rise of modern logic'. In Schilpp *1941a*: 125-63.

1966a 'Russell's ontological development'. *Jrn. Phil.* **63**: 657-67.

1986a 'Peano as logician'. *Celebrazioni in memoria di G. Peano*, Università di Torino.

Radner, M. *1975a* 'Philosophical foundations of Russell's logicism'. *Dialogue* **14**: 241-53.

Roberts, G.W. *1979a* (ed.) *Bertrand Russell memorial volume*, London: Allen & Unwin.

Robinson, R. *1950a Definition*, Oxford: Clarendon.

Rodríguez-Consuegra, F.A. 1987a *El método en la filosofía de Bertrand Russell. Un estudio sobre los orígenes de la filosofía analítica a través de la obra de Russell, sus manuscritos inéditos y los autores que más le influenciaron.* Ph. D. thesis, University of Barcelona, x + 800 pp.

1987b 'Bibliografía de Bertrand Russell en español'. *Mathesis* **3**: 183-197.

1987c 'Russell's logicist definitions of numbers 1899-1913: chronology and significance'. *Hist. Phil. Log.* **8**: 141-69.

1988a 'Bertrand Russell 1898-1900: una filosofía de la matemática inédita'. *Mathesis* **4**: 3-76.

1988b 'Elementos logicistas en la obra de Peano y su escuela'. *Mathesis* **4**: 221-299.

1988c 'Bertrand Russell 1900-1913: los principios de la matemática, parte 1ª'. *Mathesis* **4**: 355-392.

1988d 'Bertrand Russell 1900-1913: los principios de la matemática, parte 2ª'. *Mathesis* **4**: 489-521.

1989a 'Russell's theory of types, 1901-1910: its complex origins in the unpublished manuscripts'. *Hist. Phil. Log.* **10**: 131-164.

1989b 'The origins of Russell's theory of descriptions according to the unpublished manuscripts'. *Russell* **9**: 99-132.

1989c 'La "pérdida de certidumbre" en la matemática y la ciencia contemporáneas'. *Mathesis*, in print.

1990a 'Bertrand Russell 1895-1898: una filosofía prelogicista de la geometría'. *Diálogos* **55**: 71-123.

1990b 'Russell's first technical philosophy '. Essay-review of I. Winchester and K. Blackwell (eds.), *Antonomies and Paradoxes. Studies in Russell's early philosophy,* Hamilton: McMaster University Press, 1989 (*Russell* **8**, 1988, nos. 1-2). *Hist. Phil. Log.* **11**: 225-230.

1990c 'A global viewpoint on Russell's philosophy'. Essay-review of C. Wade Savage and C. Anthony Anderson (eds.), *Rereading Russell: Essays in Bertrand Russell's metaphysics and epistemology*, vol. XII of Minnesota Studies in the Philosophy of Science, Minneapolis: University of Minnesota Press, 1989. *Diálogos* **57**: 173-186.

1990d 'El impacto de Wittgenstein sobre Russell: últimos datos y vision global'. To appear.

1990e 'El logicismo russelliano: su significado filosófico'. *Crítica* **67**, in print.

1990f 'La primera filosofía de Moore'. Part I, *Agora* **9**: 153-170; part II, *Agora* **10**, in print.

1990g 'La influencia de Bradley en los orígenes de la filosofía analítica'. *Análisis filosófico*, forthcoming

1990h 'Some new light on Russell's "inextricable tangle" about meaning and denotation'. Forthcoming in *Russell*.

1990i 'A comparison among the theories of descriptions by Frege, Peano and Russell'. To appear.

1990j 'Bertrand Russell 1920-1948: una filosofía de la ciencia entre el holismo y el atomismo'. *Diálogos* **59**, forthcoming.

1990k 'La interpretación russelliana de Leibniz y el atomismo metodológico de Moore'. *Diánoia* **36**: 121-156.

1991a 'Bertrand Russel and his contemporaries'. Essay-review of E.R. Eames, *Bertrand Russell dialogue with his contemporaries*, Carbondale: Southern Illinois University Press, 1989. Forthcoming in *Russell*.

1992a 'Mathematical logic and logicism from Peano to Quine'. In I. Grattan-Guinness (ed.), *Encyclopaedia of the history and philosophy of the mathematical sciences*, London: Routledge, in print.

Sacristán, M. *1964a Introducción a la lógica y al análisis formal*, Barcelona: Ariel.

Sainsbury, R.M. *1979a Russell*, London: Routledge.

Sanzo, U. *1983a* 'Philosophie et science dans la pensée de Louis Couturat'. In Couturat *et al. 1983a*: 69-80.

Schmid, A.F. *1983a* 'La correspondance inédite entre B. Russell et L. Couturat'. *Dialectica* **37/2**: 75-108.

Schilpp, P.A. *1941a* (ed.) *The philosophy of A.N. Whitehead*, New York: Tudor, 1951.

1942a (ed.) *The philosophy of G.E. Moore*, La Salle, Ill.: Open Court, 1968.

1944a (ed.) *The philosophy of B. Russell*, La Salle, Ill.: Open Court, 1971.

Schoenman, R. *1967a* (ed.) *Bertrand Russell. Philosopher of the century*. Spanish trans. (*Homenaje a Bertrand Russell*) by C. Ulises Moulines, Barcelona: Oikos-Tau, 1968.

Smith, D.E. *1929a A source book in mathematics*, New York: Dover, 1959.

Spadoni, C. *1977a Russell's rebellion against neo-Hegelianism*. Unpublished Ph. D. thesis, University of Waterloo, Ontario, Canada.

Sprigge, T. *1979a* 'Russell and Bradley on relations'. In Roberts *1979a* (ed.): 150-70.

Stallo, J.B. *1882a The concepts and theories of modern physics*, London: Kegan Paul, 1900.

Sttebing, L.S. *1930a A modern introduction to logic*, London: Methuen. Spanish trans., Mexico: UNAM, 1965.

1942a 'Moore's influence'. In Schilpp *1942a*: 515-32.

Terracini, A. *1955a* (ed.) *In memoria di G. Peano*, Cuneo: Lic. Sci. Statale.

Thibaud, P. *1975a La logique de C.S. Peirce*, Editions de l'Université de Provence. Spanish trans., Madrid: Paraninfo, 1982.

Thomas, J.E. and Blackwell, K. *1976a* (eds.) *Russell in review*, Toronto: Hakkert.

Torretti, R. *1978a Philosophy of geometry from Riemann to Poincaré*, Dordrecht: Reidel.

Vailati, G. *1892a* 'Sui principi fondamentali della Geometria della retta'. *Rev. Mathém.* **2**: 71-75.

1899a 'La logique mathématique et sa nouvelle phase de développment dans les écrits de M. G. Peano'. *Rev. Mét. Mor.* **7**: 86-102.

van Heijenoort, J. *1967a* (ed.) *From Frege to Gödel: a source book in mathematical logic, 1879-1931*, Cambridge, Mass.: Harvard University Press.

Vuillemin, J. *1962a La philosophie de l'algèbre, I*, Paris: PUF.

1968a Leçons sur la première philosophie de Russell, Paris: Colin.

1971a La logique et le monde sensible, Paris: Flammarion.

1972a 'Platonism in Russell's early philosophy and the principle of abstraction'. In Pears *1972a*: 305-24.

1972b 'Logical flaws or philosophical problems: on Russell's *Principia mathematica*'. *Rev. Int. Phil.*

4: 534-56.

1975a 'Le "Platonisme" dans la première philosophie de Russell et le "principe d'abstraction"'. *Dialogue* **14**: 222-40.

Wang, Hao *1957a* 'The axiomatization of arithmetic'. *Jrn. Symb. Log.* **22**: 145-58.

1967a 'Russell and his logic'. *Ratio* **7**: 1-34. Spanish trans.: *Teorema* **4** (1971): 31-76; **8** (1972): 91-8.

Ward, J. *1899a Naturalism and agnosticism,* London: Black, 1903.

Watling, J. *1970a Bertrand Russell*, Edinburgh: Oliver.

Weitz, M. *1944a* 'Analysis and the unity of Russell's philosophy'. In Schilpp *1944a*: 57-121.

Weyl, H. *1949a Philosophy of mathematics and natural science*, Princeton: University Press.

Whitehead, A. N. *1898a A treatise on universal algebra,* Cambridge: University Press.

1902a 'On cardinal numbers'. *Amer. Jrn. Math.* **24**: 367-94.

Winslade, W.J. *1970a* 'Russell's theory of relations'. In Klemke *1970a*: 81-101.